Advanced Voltage Quality Controller for New Power Distribution Systems

Fei Jiang · Qi Guo · Chunming Tu

Advanced Voltage Quality Controller for New Power Distribution Systems

Springer

Fei Jiang
Changsha University of Science and Technology
Changsha, China

Qi Guo
Hunan University
Changsha, China

Chunming Tu
Hunan University
Changsha, China

ISBN 978-981-97-9860-5 ISBN 978-981-97-9861-2 (eBook)
https://doi.org/10.1007/978-981-97-9861-2

This work was supported by National Natural Science Foundation of China (52377166, 52007059).

This Springer imprint is published by the registered company Springer Nature Singapore Pte Ltd.
The registered company address is: 152 Beach Road, #21-01/04 Gateway East, Singapore 189721, Singapore

Preface

With the orderly advancement of new energy generation and the construction of the energy Internet, the trend of grid electrification becomes increasingly apparent, thereby bringing about a series of new problems and challenges in new power distribution systems. Issues such as the intermittency and uncertainty of renewable energy output make the problem of power grid energy quality more complex. The large-scale integration of power electronic loads weakens the resilience of the power consumption end against supply disturbances. The issues of grid safety, reliability, economy, and efficiency deserve continuous attention nowadays.

The series-connected voltage quality controller is connected to both ends or even multiple ends of the power grid, with a high degree of coupling. During operation, it directly affects the voltage and power flow of the grid, effectively achieving voltage fluctuation control, flexible power flow regulation, and fault overcurrent limitation. However, the effective operation time of the series-connected voltage quality controller is short, and its equipment utilization rate is low. The way the series-connected voltage quality controller is connected to the grid makes it susceptible to voltage and current surges on the grid side, posing significant reliability issues. To effectively address key foundational issues such as efficiency optimization, utilization rate improvement, and reliability enhancement of the series-connected voltage quality controller and further promote its engineering practicality, conducting theoretical and applied research on series-connected voltage quality controllers is of great significance. Due to the lack of systematic introduction to the operating principles, topology, key technologies of series-connected voltage quality controllers both domestically and internationally, this book introduces the theoretical research and application of series-connected voltage quality controllers from three aspects: optimization operation strategy, topology reconstruction technologies for functional integration, and engineering examples.

This book serves as a guideline for scholars, students, engineers, and equipment manufacturers in the fields of power electronics technology and advanced power quality control technology. We hope that this book will pave the way for enhancing the utilization and safe and efficient operation of smart grid equipment, thereby establishing a new approach and developing a new set of research and application methods for power electronic equipment, ultimately facilitating innovation in power electronic equipment technology.

Changsha, China
May 2024

Fei Jiang
Qi Guo
Chunming Tu

Acknowledgements

For the successful completion of this book, Fei Jiang is indebted to Prof. Chunming Tu for her fruitful discussions; Qi Guo for her helpful advice; and Xing Peng, Litao Xie, and Zhi Zhou for their proofreading. Furthermore, his greatest gratitude goes to his parents and grandparents for their forever love and spiritual support, and to his sister and brothers for their contributions to family.

The authors thank National Natural Science Foundation of China (Grant No. 52377166, 52007059). Meanwhile, the authors thank State Key Laboratory of Disaster Prevention and Reduction for Power Grid for providing the experimental platform.

Contents

Chapter 1
A Brief History of Voltage Quality Controllers

This chapter firstly introduces typical topology structure and evolution trend of series-connected Power electronic devices. Subsequently, the bottleneck issues faced by VQC based on series-connected power electronic were introduced. Then, the current state-of-the-art multifunctional techniques and their corresponding strategies for voltage quality controller (VQC) is introduced as well. Based on this examination, the pivotal issues and challenges encountered in the research and engineering applications of multifunctional VQC systems are succinctly summarized.

1.1 Introduction

In the contemporary era, the integration of power electronic devices into every facet of the electrical grid—spanning generation, transmission, distribution, and consumption—has surged due to their inherent attributes of intelligence and adaptability [1–3]. This proliferation has catalyzed a paradigm shift in the grid's operational landscape, concurrently ushering in a constellation of novel issues and obstacles [4, 5]. Prominent among these challenges is the intricate voltage quality quandary exacerbated by the intrinsic variability and unpredictability of renewable energy sources [6]. Additionally, the pervasive incorporation of power electronic loads has progressively diminished the resilience of end-users against supply disruptions [7]. Moreover, within this transformed setting, power flows within distribution networks have become markedly convoluted, rendering operational states more susceptible to environmental vicissitudes and the vagaries of load dynamics [8].

Power electronic devices, when series-connected, engage with the power grid at dual or even multiple points of interface [9]. Given their significant level of interconnectivity, they wield substantial influence over grid voltages and power flows, facilitating effective voltage fluctuation governance, adaptable power flow modulation, and constraint of fault-induced overcurrents [10–12]. Central to power flow

F. Jiang et al., *Advanced Voltage Quality Controller for New Power Distribution Systems*, https://doi.org/10.1007/978-981-97-9861-2_1

control, a suite of series-connected power electronic devices takes center stage. These include the Thyristor-Controlled Series Capacitor (TCSC) [13–18], Static Synchronous Series Compensator (SSSC) [19–24], Unified Power Flow Controller (UPFC) [25–30], and Interline Power Flow Controllers (IPFCs) [31–36]. An innovative approach, known as the Hybrid Power Flow Controller (HPFC), has recently been introduced, featuring a configuration of dual shunt Modular Multilevel Converters (MMCs) coupled with a series capacitor [37]. Moreover, solid-state current limiters, leveraging power electronics [38–41], have proven effective in curbing fault currents. These encompass various designs such as resonant current limiters [42, 43], bridge current limiters [44], and superconducting solid-state current limiters [45].

On the front of voltage quality compensation, the Series Active Power Filter (SAPF) [46–50], Unified Power Quality Conditioner (UPQC) [51–58], and Dynamic Voltage Restorer (DVR) are recognized as key solutions. Notably, extensive research on VQCs has been undertaken [59–63], covering areas such as topology innovation [64–70], advancement in control and optimal operational strategies [71–78], and the diversification of application contexts [79–83].

Hybrid Circuit Breakers (HCBs) [84–86] and Solid-State Circuit Breakers (SSCBs) [87–90] are likewise integrated into the grid in a series configuration, distinguished by their low on-state dissipation. These devices can swiftly isolate faulty circuits or segments, thereby mitigating potential damage and enhancing system safety. Furthermore, the Electrical Spring (ES) [91–93], a dynamic load management mechanism, and the series-connected impedance measurement unit [94] also fall under the category of series-connected power electronic devices. These components serve critical roles in monitoring and controlling grid conditions. Additionally, the series-type damper [95, 96], designed to stabilize power systems by dampening oscillations, and the secondary ripple suppression device for the DC link of converters [97], aimed at smoothing out voltage fluctuations, further expand the capabilities of series-connected devices in power electronics. The basic classification of series-connected power electronic devices is shown in Fig. 1.1.

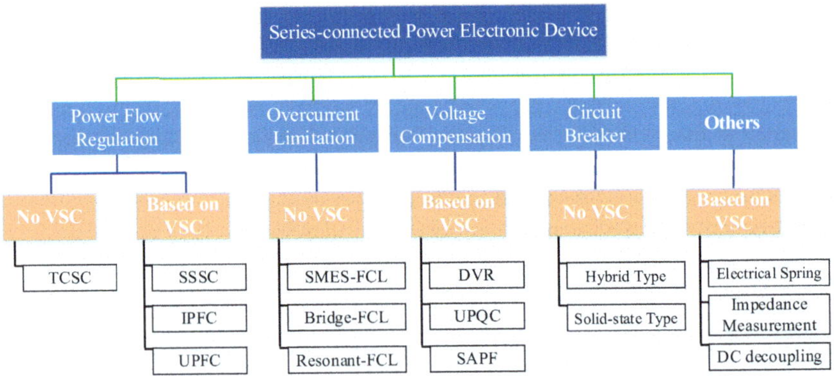

Fig. 1.1 Classification of series-connected power electronic devices

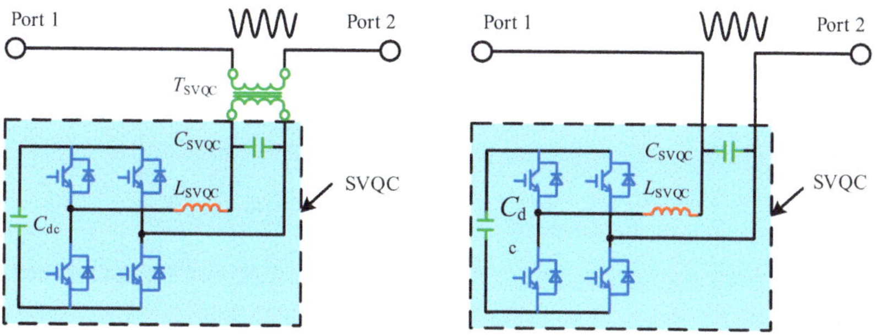

Fig. 1.2 Typical access scheme of the SVQC

Among the plethora of series-connected power electronic apparatuses, the series-connected voltage quality controller (SVQC) typically interfaces with the power grid through a filter capacitor or a transformer [98–103], depicted schematically in Fig. 1.2a and b. By modulating the magnitude and phase angle of its output voltage, the SVQC facilitates agile power flow regulation and terminal voltage control. Contrasted with conventional power flow control mechanisms like the TCSC, the UPFC predicated on SVQC boasts superior adaptability and diminished harmonic distortion. Similarly, juxtaposed against traditional voltage quality control devices such as On-Load Tap Changers (OLTCs), the SVQC-based VQC or UPQC can execute precise compensation through continuous adjustment. However, the SVQC is not without drawbacks; its reliance on power semiconductor components can lead to limitations in electrical stress endurance and elevated costs. Notwithstanding these challenges, the ongoing evolution of power electronics technology portends a promising future for SVQC, positioning it as a leading trend in the advancement of series-connected devices.

It is crucial to highlight that, despite initial hurdles, the maturation of power electronics is steadily amplifying the benefits and applicability of SVQC, underscoring its potential to revolutionize power system control and reliability.

Moreover, in comparison with the PVQC (Fig. 1.3), the SVQC is distinguished by its connection at both extremities, or indeed multiple points, across the power grid [9]. This unique configuration enables the SVQC to exert a direct influence on grid voltages and power flows, thereby facilitating efficacious voltage quality management, adaptive power flow modulation, and the imposition of limits on fault overcurrents.

The SVQC's strategic placement within the network allows it to intervene dynamically in the electrical parameters of the grid, making it an effective tool for voltage regulation and power distribution optimization. Its capability to limit overcurrents during faults also enhances system resilience, contributing to improved overall grid stability and reliability. The main applications and function performance comparison between SVQC and PVQC is shown in Table. 1.1.

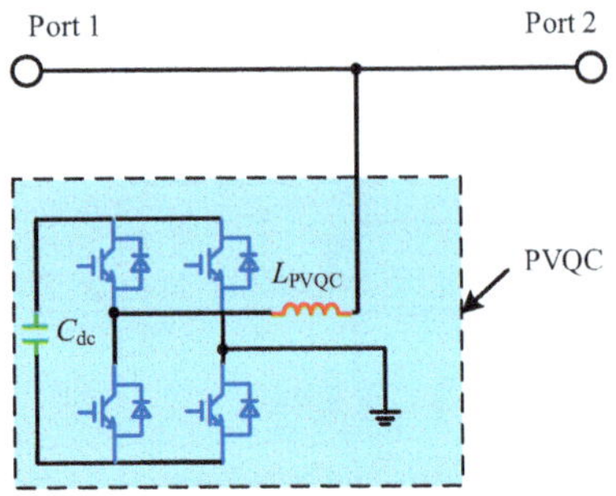

Fig. 1.3 Typical access scheme of the PVQC

Table 1.1 Main applications and function performance comparison between SVQC and PVQC

Function	Performance of SVQC	Performance of PVQC
Voltage fluctuation control	Good	Poor
Voltage harmonic control	Good	/
Fault overcurrent limitation	Good	/
Power flow regulation	Good	Poor
Current harmonic control	/	Good
Reactive power compensation	Poor	Good
Gird integration of renewable energy generation	Poor	Good

However, the uniqueness of the connection methodology for the SVQC, while imparting distinct performance perks, simultaneously ushers in substantial practical impediments. Predominant among these challenges are:

Brief active engagement: given that SVQCs predominantly spring into action under anomalous grid scenarios, the active operational span of a typical SVQC tends to be limited, leading to a suboptimal utilization quotient. Events such as voltage perturbations and grid anomalies, although detrimental to both the load and grid integrity, manifest infrequently and are transient in nature. This results in a low duty cycle for series-connected power electronic devices, including solid-state fault current limiters and dynamic voltage restorers, which remain largely idle, incurring standby losses.

Resilience against grid transients: despite its inherent susceptibility, the SVQC must consistently weather the storm of voltage and current surges commonplace along the grid lines. Its intimate connection with the grid means that any shifts in grid operational dynamics can have a cascading effect on the SVQC's output performance and stability, potentially amplifying voltage and current instabilities onto the SVQC itself.

These considerations highlight the intricate balance between the SVQC's operational efficacy and its practical viability, necessitating robust design and control strategies to mitigate the impact of these challenges.

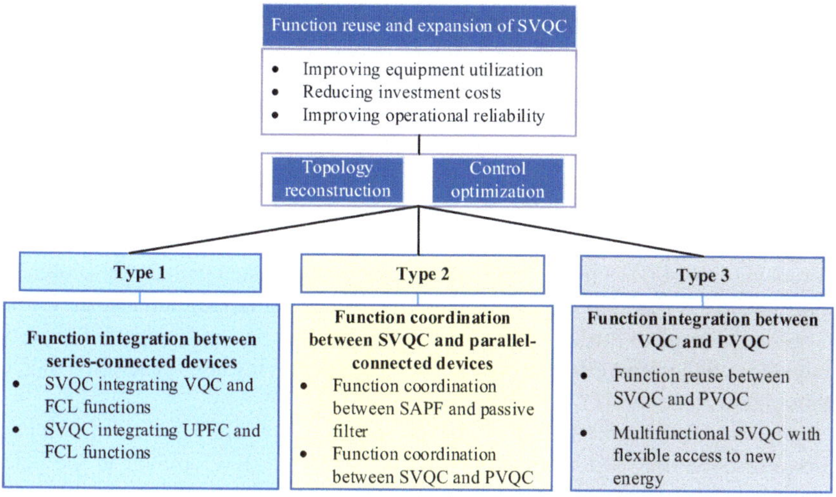

Fig. 1.4 Research direction of function integration and coordination of the existing SVQC

In view of above challenges in SVQC, multifunctional operation is an effective way for SVQC to break through the bottleneck problems, such as single function and low equipment utilization. The multifunctional operation means to endow SVQC with multiple functions by means of topology reconstruction or control optimization, as shown in Fig. 1.4, which can include the two or more functions of series-connected power electronic device according to the application occasions; or makes SVQC assist parallel-connected devices to optimize the operation performance; or changes grid connection mode of SVQC to achieve the functions of parallel-connected devices, such as new energy output and reactive power compensation. The solutions mentioned above can greatly improve the application value of SVQC, as follows.

The function integration between different series-connected power electronic devices: Superimpose the fundamental wave component and harmonic component in the control signal, so that the SVQC has the functions of the VQC and SAPF, thus effectively ensuring both the high reliability and quality of the power supply. Moreover, relevant scholars integrate the fault current-limiting function into the SVQC with voltage compensation or power flow control ability through topology reconstruction or control optimization, so that the SVQC can achieve fault ride-through.

The function coordination between the SVQC and parallel-connected power electronic devices: Such function coordination can substantially improve the harmonic control and reactive power compensation capabilities of parallel-connected passive devices and even reduce the capacity demand of parallel-connected power devices.

Flexible switching of functions between the SVQC and PVQC: Some scholars have realized the access mode transformation of the VSC between series-connection and parallel-connection through the solid-state switch or relay. The device works in the harmonic control/renewable energy grid-connected mode when the grid is normal and switches to the voltage compensation mode when the grid is disturbed, which effectively improves the utilization and operating efficiency of the SVQC.

Exploring the pivotal challenges and innovations in enhancing SVQC capabilities offers a promising avenue to augment the utilization efficiency of smart grid assets and bolster operational effectiveness. The core objective of this review is to encapsulate the advancements in extending SVQC functionalities. It delves into the theoretical foundations and practical methodologies associated with the development of multifaceted SVQC topologies and control enhancements. Our aim is to galvanize interest and stimulate further investigation among academic and industrial experts, catalyzing collaborative efforts toward refining and deploying next-generation SVQC technologies.

By synthesizing the latest findings and perspectives, we aspire to create a comprehensive resource that will inspire ongoing research and development in the field, ultimately contributing to the realization of smarter, more resilient, and efficient electrical grids. This endeavor is crucial for addressing the evolving demands of modern power systems, particularly in the context of integrating renewable energy sources and managing complex grid dynamics.

1.2 Function Integration Method Between Series-Connected Power Electronic Devices

PVQCs demonstrate versatility by manipulating various reference current components—active, reactive, and harmonic—to meet diverse control objectives, as detailed in literature [104, 105]. Analogously, SVQCs exhibit flexibility in control across two primary dimensions: power frequency components and harmonic components. Regarding power frequency voltage and current control, SVQCs can accomplish tasks such as mitigating voltage sags and swells, regulating power flows, and limiting fault currents during line anomalies. For harmonic component control, they are capable of compensating for voltage harmonics. Multifunctional operation, therefore, signifies equipping SVQCs with a combination of these aforementioned capabilities tailored to specific application contexts. Primarily, multifunctional SVQCs can concurrently manage fundamental and harmonic components to serve as both VQCs and Static Active Power Filters (SAPFs), ensuring enhanced power supply reliability and quality. These concepts have been extensively discussed in [106, 107] and will not be reiterated here. Moreover, considering the essential requirements for power quality and grid safety, SVQCs integrated with fault current limitation capabilities represent a strategic advancement in boosting device utility and safeguarding both the apparatus and the grid proactively. The implementation of fault current limitation in SVQCs is predominantly achieved through optimized control strategies and enhanced topological designs. A summary of the highest number of

Table 1.2 Typical multifunctional method with different series-connected power electronic devices integration

Type	Functions	Maximum number of functions
(1) VQC + SAPF	Voltage sag/swell compensation Voltage harmonic control	2
(2) UPQC/VQC + FCL	Voltage sag/swell compensation Voltage harmonic control Line fault current limiting	3
(3) UPFC + FCL	Voltage sag/swell compensation Power flow regulation Line fault current limiting	3

objectives accomplished by representative multifunctional approaches is delineated in Table 1.2.

1.2.1 The SVQC Integrated the Function of Fault Current Limiter Based on Topology Improvement

The conceptual framework behind a multifunctional SVQC endowed with fault current limitation capabilities hinges on a meticulous examination of the converter's port characteristics and topological alterations in response to a short-circuit event. Through either refining the control strategy of the converter system or leveraging shared core components, a substantial impedance barrier is established, effectively capping the surge in fault-line currents (Table 1.3).

Literature [113–117] introduces a hybrid system that amalgamates a Fault Current Limiter (FCL) module with a UPFC, coined as FCL-UPFC. This system builds upon a conventional UPFC topology by incorporating an FCL module, centered around a bridge circuit, to create a protective layer. The FCL module is strategically connected to the secondary side of the series transformer, as depicted in Fig. 1.2. This arrangement ensures that the SVQC is shielded from the direct impact of elevated currents and voltages during system disturbances.

Subsequent enhancements [115–117] have sequentially emerged, illustrated as "Structures 1, 2, and 3" in Fig. 1.5. These iterations of the FCL-UPFC seamlessly merge power flow regulation and power quality enhancement under normal grid conditions while seamlessly transitioning to fault current limitation during line short-circuits. This exemplifies an SVQC proficient in fault current limitation, showcasing commendable potential for widespread adoption. Nevertheless, the functional modules within these configurations operate with a degree of autonomy, suggesting room for optimization in the integration efficiency of individual components. Further refinements could enhance the synergy and multifunctionality of the system, leading to a more compact and versatile solution.

Table 1.3 Comparison of function integration schemes of different series power electronic devices

Scheme		Advantage	Disadvantage
Topology reconstruction	Simple combination of a VQC and fault current limiter [108–112]	(1) Simple topology	(1) Low utilization of components
	FCL-UPFC [113–117]	(1) Limiting the fault current to 0	(1) Low utilization of components (2) The energy loss during FCL mode cannot be avoided
Control optimization	System reverse voltage compensation [118]	(1)No need to add auxiliary components	(1) Equipped with large capacity energy storage on DC link side (2) High-power rating level of converter
	VQC with energy storage based on fixed/variable virtual inductance [119–124]	(1) No need to add auxiliary components (2) No active power loss	(1) High-power rating level of converter
	Back-to-back VQC-based on variable virtual impedance [125]	(1) No need to add auxiliary components (2) No active power loss	(1) High-power rating of converter (2) Complex topology and a high cost

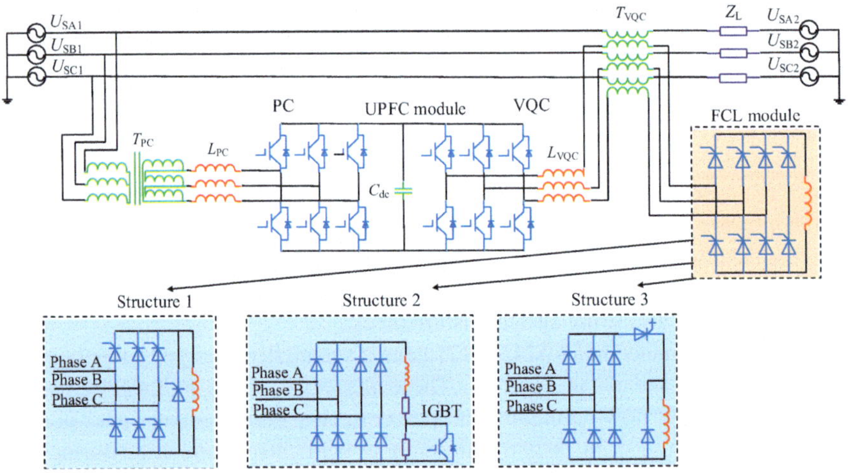

Fig. 1.5 Topology of the FCL-UPFC and its improved topology

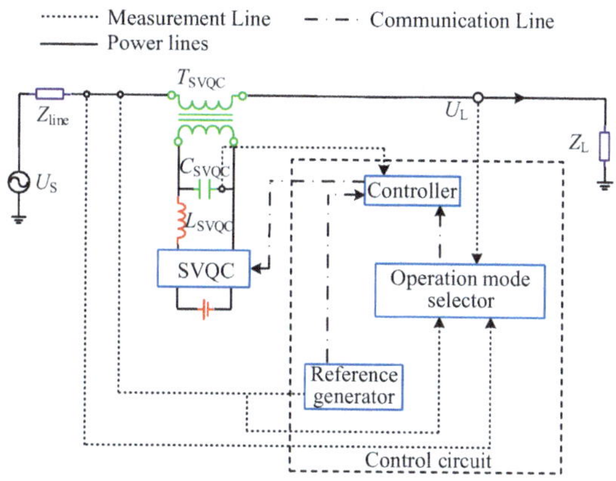

Fig. 1.6 Improved control block diagram of the SVQC with a current-limiting function

1.2.2 The SVQC Integrated the Function of Fault Current Limiter Based on Control Optimization

By improving the basic operation strategy of the SVQC, the equivalent output impedance of the converter can be effectively controlled, thereby realizing the power quality adjustment and the fault current limitation.

In reference [118], an innovative strategy is employed, where the SVQC delivers a compensatory voltage that is phase-opposed to the grid voltage in the event of a short-circuit fault. This action results in the load-side voltage being virtually nil, thereby effectively curtailing the fault-induced overcurrent. The principal control schematic, delineated within dashed lines in Fig. 1.6, illustrates this mechanism.

However, given that this approach necessitates the SVQC to generate a significant volume of active power, it places stringent demands on the DC link capacity. These demands are particularly challenging to meet in medium- to high-voltage and high-capacity systems, making the strategy less feasible in such applications.

To address these limitations, refinements were introduced in reference [119]. These optimizations led to a notable reduction in the DC link energy requirement for the SVQC, thereby enhancing its performance in terms of current limitation and the transition between operational modes. This improved methodology proved more viable and efficient, especially concerning the current-limiting efficacy and the agility of mode transitions, making it a superior alternative for broader system integration.

A simulation of a virtual impedance connection is achieved through the strategic design of an RL feedforward algorithm aimed at curbing overcurrents during faults, a concept visualized in Fig. 1.7. This methodology effectively harnesses the principles of resistance and inductance to manage and limit current surges under fault conditions.

In [122], the perception of equivalent impedance specific to the VQC is realized by engineering a flux linkage. This approach ensures the cessation of reactive power exchange throughout the fault duration, showcasing an advanced feature in

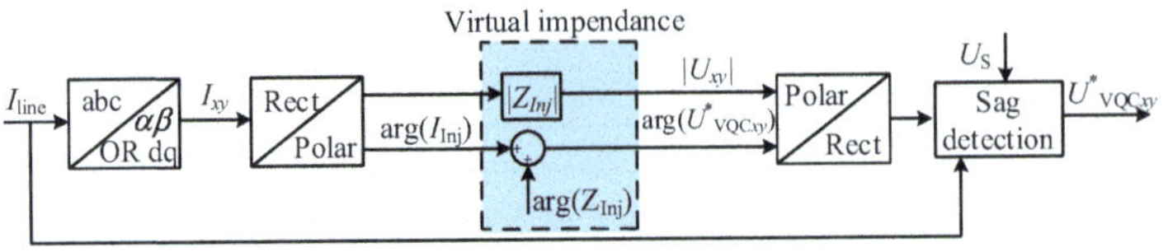

Fig. 1.7 RL feedforward current-limiting algorithm

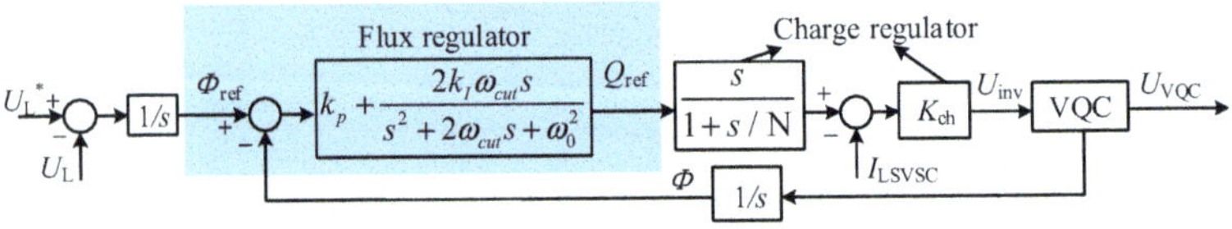

Fig. 1.8 Improved flux control algorithm

the damping of current surges, as demonstrated in Fig. 1.8. This technique provides a superior means of controlling current flow during critical events.

Building upon these foundational insights, an enhancement to the virtual impedance control strategy is presented in [123].

Where Φ is the flux, Q_{ref} is reference charge, K_P is the proportional, K_I is the integral term for fundamental frequency ($\omega_0 = 2\pi f$ rad/s, 50 Hz), N is the regulator gain, K_{ch} is the proportional of derivative compensator, and ω_{cut} is the cutoff bandwidth for adjusting the controller performance for the frequency variations.

To augment the operational efficacy of the SVQC, as detailed in [124], the VQC is engineered to emulate virtual capacitance characteristics, enabling it to counteract voltage sags on the line. Moreover, in scenarios involving short-circuit faults on the load side, the VQC adopts virtual inductance properties to cap overcurrents. These stratagems ensure that the VQC can nimbly switch between operational modes contingent on the prevailing grid conditions. However, the complexity of the control strategy warrants a more thorough exploration of the transitional dynamics when shifting between distinct modes. This involves scrutinizing the transient processes to refine the switching mechanisms for smoother transitions.

Furthermore, the methodology for calculating voltage amplitude and phase angle, as put forth in [125], enables rapid computation of the fundamental voltage signal. This serves as a critical reference point for controlling the system during periods of fault current limitation, ensuring that the system remains insulated from overcurrent stresses. An additional benefit is the minimal loss of active power during the current limitation phase.

In essence, fault current limitation schemes predicated on topological modifications often suffer from reduced topological integration and diminished equipment utilization efficiency. Conversely, strategies centered on control optimization may demand substantial energy backup or necessitate converters with higher power ratings. Therefore, to satisfy the pragmatic requisites of engineering applications, the amalgamation of SVQC with fault current-limiting functionalities necessitates further research and refinement. This pursuit aims to strike an optimal balance between topological simplicity, control sophistication, and operational efficiency, thereby catering to the stringent requirements of real-world deployments.

1.3 Challenges and Future Prospects

The multifunctional SVQC has garnered considerable research momentum, signifying a pivotal evolutionary trajectory for SVQCs. Nevertheless, the widespread adoption and practical application of this technology face numerous hurdles that necessitate urgent attention and innovative breakthroughs. The subsequent discussion delves into the current limitations in research and anticipates future directions for multifunctional SVQCs, structured around four core aspects: advancements in power semiconductor devices, innovations in topological structures, refinements in operational strategies, and progress in engineering applications. The details are as follows:

(1) Topology optimization of the multifunctional SVQC

Currently, the approach to achieving functional reuse in series and series/parallel converters through physical switches often entails a heightened topological complexity, accompanied by notable voltage and current transients during the switching process [43–46]. When relying on control optimization strategies for functional reuse, the distinct active and reactive output demands of SVQC under varying operational modes pose significant challenges for comprehensive optimization of converter capacity and the dc link side [47, 48]. Furthermore, devising a multifunctional SVQC topology that balances complexity, operational reliability, and capacity optimization remains elusive. Consequently, the future development trajectory of multifunctional SVQC necessitates achieving functional integration and expansion while minimizing the addition of physical switches and topological intricacy.

(2) Operation optimization of the multifunctional SVQC

The multifunctional SVQC displays vastly different output characteristics depending on its control objectives. Switching between series and parallel access modes, or even between distinct series access modes, necessitates precise instantaneous energy balancing, output amplitude/phase transitions, and precise timing of switching device activations. Achieving seamless mode transitions poses a significant challenge and is crucial for ensuring the long-term safe and efficient operation of the multifunctional SVQC system.

References

1. B.K. Bose, Global energy scenario and impact of power electronics in 21st century. IEEE Trans. Ind. Electron. **60**(7), 2638–2650 (2016)

2. F. Blaabjerg, Y. Yang, D. Yang, X. Wang, Distributed power-generation systems and protection. Proc. IEEE. **105**(7), 1311–1321 (2017)
3. S.K. Sahoo, A.K. Sinha, N.K. Kishore, Control techniques in AC, DC, and hybrid AC–DC microgrid: a review. IEEE J. Emerg. Select. Topics Power Electron. **6**(2), 738–759 (2018)
4. Z. Zeng, H. Yang, R. Zhao, C. Cheng, Topologies and control strategies of multi-functional grid-connected inverters for power quality enhancement: a comprehensive review. Renew. Sustain. Energy Rev. **24**, 223–270 (2013)
5. V. Kharrazi, Sreeram, Y. Mishra, Assessment techniques of the impact of grid-tied rooftop photovoltaic generation on the power quality of low voltage distribution network-A review. Renew. Sustain. Energy Rev. **120**, 1–16 (2020)
6. M.S. Ali, M.M. Haque, P. Wolfs, A review of topological ordering based voltage rise mitigation methods for LV distribution networks with high levels of photovoltaic penetration. Renew. Sustain. Energy Rev. **103**, 463–476 (2019)
7. B. Singh, A. Chandra, K. Al-Haddad, *Power Quality Problems and Mitigation Techniques* (Wiley, Chichester, U.K., 2015)
8. S.M. Imdadullah, M.S. Amrr, J. Asghar, I. Ashraf, M. Meraj, A comprehensive review of power flow controllers in interconnected power system networks. IEEE Access. **8**, 18036–18063 (2020)
9. J. Roldan-Pérez, A. García-Cerrada, M. Ochoa-Gimenez, J.L. Zamora-Macho, On the power flow limits and control in series-connected custom power devices. IEEE Trans. Power Electron. **31**(10), 7328–7338 (2016)
10. Y.L. Tan, Analysis of series and shunt line compensation by a unified power flow controller. IEEE Power Eng. Rev. **19**(9), 63–645 (1999)
11. M. Mansoor, N. Mariun, A. Toudeshki, N.I.A. Wahab, A.U. Mian, M. Hojabri, Innovating problem solving in power quality devices: a survey based on dynamic voltage restorer case (DVR). Renew. Sustain. Energy Rev. **70**, 1207–1216 (2017)
12. H. Sahebi, Samet, T. Ghanbari, Evaluation of power transformer inrush currents and internal faults discrimination methods in presence of fault current limiter. Renew. Sustain. Energy Rev. **68**, 102–112 (2017)
13. Asle, M. Alizadeh, M.E. Tayeba, R. Razzaghi, B. Mohammadi-Ivatloo, An adaptive cumulative sum based method for unblocking distance relays in TCSC compensated transmission lines. Int. J. Elec. Power Energy Syst. **131**, 107095 (2021)
14. L. Chang, Y. Liu, Y. Jing, X. Chen, J. Qiu, Semi-globally practical finite-time H_∞ control of TCSC model of power systems based on dynamic surface control. IEEE Access. **8**, 10061–10069 (2020)
15. D.D. Rosso, C.A. Canizares, V.M. Dona, A study of TCSC controller design for power system stability improvement. IEEE Trans. Power Syt. **18**(4), 1487–1496 (2003)
16. M. Khederzadeh, T. Sidhu, Impact of TCSC on the protection of transmission lines. IEEE Trans. Power Del. **21**(1), 80–87 (2006)
17. S. He, J. Suonan, Z.Q. Bo, Integrated impedance-based pilot protection scheme for the TCSC-compensated EHV/UHV transmission lines. IEEE Trans. Power Del. **28**(2), 835–844 (2013)
18. M. Esmaili, M. Ghamsari-Yazdel, N. Amjady, C.Y. Chung, A.J. Conejo, Transmission expansion planning including TCSCs and SFCLs: a MINLP approach. IEEE Trans. Power Syst. (2020) https://doi.org/10.1109/TPWRS.2020.2987982
19. L. Gyugyi, C.D. Schauder, K.K. Sen, Static synchronous series compensator: a solid-state approach to the series compensation of transmission lines. IEEE Trans. Power Del. **12**(1), 406–417 (1997)
20. F.A.R.A. Jowder, B.T. Ooi, Series compensation of radial power system by a combination of SSSC and dielectric capacitors. IEEE Trans. Power Del. **20**(1), 458–465 (2005)
21. F. Mancilla-David, S. Bhattacharya, G. Venkataramanan, A comparative evaluation of series power-flow controllers using DC- and AC-Link converters. IEEE Trans. Power Del. **23**(2), 985–996 (2008)
22. J. Morsali, K. Zare, M.T. Hagh, A novel dynamic model and control approach for SSSC to contribute effectively in AGC of a deregulated power system. Int J Elec Power Energy Syst. **95**, 239–253 (2018)

23. R. Thirumalaivasan, M. Janaki, N. Prabhu, Damping of SSR using subsynchronous current suppressor with SSSC. IEEE Trans. Power Del. **28**(1), 64–74 (2013)
24. A.H. Movahedia, G.B. Niasara, Gharehpetianb, Designing SSSC, TCSC, and STATCOM controllers using AVURPSO, GSA, and GA for transient stability improvement of a multi-machine power system with PV and wind farms. Int. J. Elec. Power Energy Syst. **106**, 455–466 (2019)
25. L. Liu, P. Zhu, Y. Kang, J. Chen, Power-flow control performance analysis of a unified power-flow controller in a novel control scheme. IEEE Trans. Power Del. **22**(3), 1613–1619 (2007)
26. P. Singh, N. Senroy, Steady-state models of STATCOM and UPFC using flexible holomorphic embedding. Electr. Power Syst. Res **199**, 107390 (2021)
27. Y. Liu, S. Yang, X. Wang, D. Gunasekaran, U. Karki, F.Z. Peng, Application of transformer-less UPFC for interconnecting two synchronous AC grids with large phase difference. IEEE Trans. Power Electron. **31**(9), 6092–6103 (2016)
28. H. Li, T. Zheng, S. Huang, Y. Wang, UPFC fault ride-through strategy based on virtual impedance and current limiting reactor. Int. J. Elec. Power Energy Syst. **125**, 106491 (2021)
29. Z. Moravej, M. Pazoki, M. Khederzadeh, New smart fault locator in compensated line with UPFC. Int. J. Elec. Power Energy Syst. **92**, 125–135 (2017)
30. F.H. Gandoman, A. Ahmadi, A.M. Sharaf, P. Siano, J. Pou, B. Hredzak, V.G. Agelidis, Review of FACTS technologies and applications for power quality in smart grids with renewable energy systems. Renew. Sustain. Energy Rev. **82**, 502–514 (2018)
31. P. Singh, N. Senroy, Steady-state model of VSC based FACTS devices using flexible holomorphic embedding: (SSSC and IPFC). Int. J. Elec. Power Energy Syst. **133**, 107256 (2021)
32. K.R. Padiyar, N. Prabhu, Analysis of SSR with three-level twelve-pulse VSC-based interline power flow controller. IEEE Trans. Power Del. **22**(3), 1688–1695 (2007)
33. M.R. Khodabakhshian, M. Esmaili, Bornapour, Optimal coordinated design of UPFC and PSS for improving power system performance by using multi-objective water cycle algorithm. Int. J. Elec. Power Energy Syst. **83**, 124–133 (2016)
34. P. Singh, N. Senroy, R. Tiwari, Guaranteed convergence embedded system for SSSC and IPFC. IEEE Trans. Power Syst. **36**(3), 2725–2728 (2021)
35. R. Vasquez-Arnez, L. Zanetta, A novel approach for modeling the steady-state VSC-based multiline FACTS controllers and their operational constraints. IEEE Trans. Power Del. **23**(1), 457–464 (2008)
36. M. Ebeed, S. Kamel, F. Jurado, Determination of IPFC operating constraints in power flow analysis. Int. J. Electr. Power Energy Syst. **81**, 299–307 (2016)
37. M.S., X. Meng, L. Wang, B. Mohammadi-Ivatloo, A decentralized multi-Loop scheme for robust control of a power flow controller with two shunt modular multilevel converters. IEEE Trans. Ind. Inf. **14**(10), 4309–4321 (2018)
38. Abramovitz, K.M. Smedley, Survey of solid-state fault current limiters. IEEE Trans. Power Electron. **27**(6), 2700–2782 (2012)
39. S. Zhang, G. Zou, X. Wei, C. Zhang, Bridge-type multiport fault current limiter for applications in MTDC grids. IEEE Trans. Ind. Electron (2021). https://doi.org/10.1109/TIE.2021.3100996
40. H. Zhou, J. Yuan, F. Chen, B. Chen, Inductive fault current limiters in VSC-HVDC systems: a review. IEEE Access. **8**, 38185–38197 (2020)
41. H. Heidarya, K. Radmaneshb, A. Rouzbehic, G.B. Mehrizi-Sanid, Gharehpetiane, Inductive fault current limiters: a review. Electr. Power Syst. Res. **187**, 106499 (2020)
42. S.B. Naderi, M. Jafari, M.T. Hagh, Parallel-resonance-type fault current limiter. IEEE Trans. Ind. Electron. **60**(7), 2538–2546 (2016)
43. M.M. Lanes, H.A.C. Braga, P.G. Barbosa, Fault current limiter based on resonant circuit controlled by power semiconductor devices. IEEE Latin. Amer. Trans. **5**(5), 311–320 (2007)
44. D. Baimel, N. Chowdhury, J. Belikov, Y. Levron, New type of bridge fault current limiter with reduced power losses for transient stability improvement of DFIG wind farm. Electric. Power Syst. Res **197**, 107293 (2021)

45. Ngamroo, An optimization of superconducting coil installed in an HVDC-wind farm for alleviating power fluctuation and limiting fault current. IEEE Trans. Appl. Supercond. **29**(2) (2019)
46. Singh, K. Al-Haddad, A. Chandra, A review of active filters for power quality improvement. IEEE Trans. Ind. Electron. **46**(5), 960–971 (1999)
47. S. Senturk, A.M. Hava, Performance enhancement of the single-phase series active filter by employing the load voltage waveform re-construction and line current sampling delay reduction methods. IEEE Trans. Power Electron. **26**(8), 2210–2220 (2011)
48. Javadi, A. Hamadi, L Woodward, K. Al-Haddad, Experimental investigation on a hybrid series active power compensator to improve power quality of typical households. IEEE Trans. Ind. Electron. **63**(8), 4849–4854 (2016)
49. L.F.J. Meloni, F.L. Tofoli, Â.J.J. Rezek, E.R. Ribeiro, Modeling and experimental validation of a single-phase series active power filter for harmonic voltage reduction. IEEE Access. **7**, 151971–151984 (2019)
50. G. Buticchi, D. Barater, C. Concari, G. Franceschini, Single-phase series active power filter with transformer-coupled matrix converter. IET Power Electron. **9**(6), 1279–1289 (2016)
51. R. Yang, J. Jin, Unified power quality conditioner with advanced dual control for performance improvement of DFIG-based wind farm. IEEE Trans. Sustain. Energ. **12**(1), 116–126 (2021)
52. H. Fujita, H. Akagi, The unified power quality conditioner: the integration of series and shunt-active filters. IEEE Trans. Power Electron. **13**(2), 315–322 (1998)
53. L. Meng, L. Ma, W. Zhu, H. Yan, T. Wang, W. Mao, X. He, Z. Shu, Controlstrategy of single-phase UPQC for suppressing the influences of low-frequency DC-Link voltage ripple. IEEE Trans. Power Electron. (2021). https://doi.org/10.1109/TPEL.2021.3106049
54. J. Jin, R. Yang, R. Zhang, Y. Fan, Q. Xie, X. Chen, Combined low voltage ride through and power smoothing control for DFIG/PMSG hybrid wind energy conversion system employing a SMES-based AC-DC unified power quality conditioner. Int. J. Elec. Power Energy Syst. **128**, 106733 (2021)
55. K. Jindal, A. Ghosh, A. Joshi, Interline unified power quality conditioner. IEEE Trans. Power Del. **22**(1), 364–372 (2007)
56. H.R. Mohammadi, R.Y. Varjani, H. Mokhtari, Multi-converter unified power-quality conditioning system: MC-UPQC. IEEE Trans. Power Del. **24**(3), 1679–1686 (2009)
57. K.H. Kwan, Y.C. Chu, P.L. So, Model-based H_∞ control of a unified power quality conditioner. IEEE Trans. Ind. Electron. **56**(7), 2493–2504 (2009)
58. M.A. Amini, A. Jalilian, Optimal sizing and location of open-UPQC in distribution networks considering load growth. Int. J. Elec. Power Energy Syst. **130**, 106893 (2021)
59. V. Khadkikar, Enhancing electric power quality using UPQC: a comprehensive overview. IEEE Trans. Power Electron. **27**(5), 2284–2297 (2012)
60. M.H.J. Bollen, *Understanding Power Quality Problems: Voltage Sags and Interruptions* (IEEE Press, New York, 2000)
61. F.H. Gandomana, A. Ahmadib, A.M. Sharafc, P. Sianod, J. Poub, B. Hredzakb, V.G. Agelidis, Review of FACTS technologies and applications for power quality in smart grids with renewable energy systems. Renew. Sustain. Energy Rev. **82**, 502–514 (2018)
62. F.G. Montoya, A. Garcia-Cruza, M.G. Montoya, F. Manzano-Asucliaroa, Power quality techniques research worldwide: a review. Renew. Sustain. Energy Rev. **54**, 846–856 (2016)
63. P. Mahela, A.G. Shaik, Topological aspects of power quality improvement techniques: a comprehensive overview. Renew. Sustain. Energy Rev. **58**, 1129–1142 (2016)
64. J. Kaniewski, P. Szczesniak, M. Jarnut, G. Benysek, Hybrid voltage sagVswell compensators: a review of hybrid ACVAC converters SAPF. IEEE Ind. Electron. Mag. **9**(4), 37–48 (2015)
65. J.G. Nielsen, F. Blaabjerg, A detailed comparison of system topologies for dynamic voltage restorers. IEEE Trans. Ind. Appl. **41**(5), 1272–1280 (2005)
66. Y. Goharrizi, S.H. Hosseini, M. Sabahi, G.B. Gharehpetian, Three-phase HFL-DVR with independently controlled phases. IEEE Trans. Power Electron. **27**(4), 1706–1713 (2012)
67. F.B. Moreno, J.E. Palacios, J. Posada, J.A. Lopez, Implementation and evaluation of a new DVR topology with AC link for series compensation. Electr. Power Syst. Res. **181**, 106184 (2020)

68. J. Wang, Y. Xing, H. Wu, T. Yang, A novel dual-DC-port dynamic voltage restorer with reduced-rating integrated DC-DC converter for wide-range voltage sag compensation. IEEE Trans. Power Electron. **34**(8), 7437–7449 (2018)
69. Ghosh, A.K. Jindal, A. Joshi, Design of a capacitor-supported dynamic voltage restorer (DVR) for unbalanced and distorted loads. IEEE Trans. Power Del. **19**(1), 405–413 (2004)
70. S. Jothibasu, M.K. Mishra, An improved direct AC–AC converter for voltage sag mitigation. IEEE Trans. Ind. Electron. **62**(1), 21–29 (2015)
71. J. Cao, H. Liu, P. Ding, B. Yang, S. Xie, Wide correction range three-level dynamic voltage corrector. IEEE Trans. Power Electron. **31**(9), 6217–6226 (2016)
72. K.A. Hosani, T.H. Nguyen, N.A. Sayari, An improved control strategy of 3P4W DVR systems under unbalanced and distorted voltage conditions. Int. J. Elec. Power Energy Syst. **98**, 233–242 (2018)
73. E. Ebrahimzadeh, S. Farhangi, H. Iman-Eini, F.B. Ajaei, R. Iravani, Improved phasor estimation method for dynamic voltage restorer applications. IEEE Trans. Power Del. **30**(3), 1467–1477 (2015)
74. S. Majumder, S.A. Khaparde, A.P. Agalgaonkar, S.V. Kulkarni, S. Perera, Graph theory based voltage sag mitigation cluster formation utilizing dynamic voltage restorers in radial distribution networks. IEEE Trans. Power Del. (2021). https://doi.org/10.1109/TPWRD.2020.3047811
75. P. Torres, P. Roncero-Sãnchez´, V.F. Batlle, A two degrees of freedom resonant control scheme for voltage-sag compensation in dynamic voltage restorers. IEEE Trans. Power Electron. **33**(6), 4852–4868 (2018)
76. F.B. Ajaei, S. Afsharnia, A. Kahrobaeian, S. Farhangi, A fast and effective control scheme for the dynamic voltage restorer. IEEE Trans. Power Del. **26**(4), 2398–2406 (2010)
77. H. Lei, X. Lin, M. Xue, Y. Kang, A discrete state-space control scheme for dynamic voltage restorers. J. Power Electron. **13**(3), 400–408 (2013)
78. S. Biricik, H. Komurcugil, N.D. Tuyen, M. Basu, Protection of sensitive loads using sliding mode controlled three-phase DVR with adaptive notch filter. IEEE Trans. Ind. Electron. **66**(7), 5465–5475 (2019)
79. M. Rauf, V. Khadkikar, An enhanced voltage sag compensation scheme for dynamic voltage restorer. IEEE Trans. Ind. Electron. **62**(5), 2683–2691 (2015)
80. A.K. Jindal, A. Ghosh, A. Joshi, Technology assessment for power quality mitigation devices—Micro-DVR case study. Electr. Power Syst. Res. **78**(2), 255–263 (2008)
81. M. Vilathgamuwa, H.M. Wijekoon, S.S. Choi, A novel technique to compensate voltage sags in multiline distribution system—the interline dynamic voltage restorer. IEEE Trans. Ind. Electron. **53**(5), 1603–1611 (2006)
82. M. Babaei, F Kangarlu, Sensitive load voltage compensation against voltage sags/swells and harmonics in the grid voltage and limit downstream fault currents using DVR. Electr. Power Syst. Res. **83**(1), 80–90 (2012)
83. M. Ramasamy, S. Thangavel, Experimental verification of PV based dynamic voltage restorer (PV-DVR) with significant energy conservation. Int. J. Elec. Power Energy Syst. **49** 296–307 (2013)
84. S.I. Gkavanoudis, C.S. Demoulias, Fault ride-through capability of a DFIG in isolated grids employing DVR and supercapacitor energy storage. Int. J. Elec. Power Energy Syst. **49**, 356–363 (2015)
85. Shukla, G.D. Demetriades, A survey on hybrid circuit-breaker topologies. IEEE Trans. Power Del. **30**(2), 627–641 (2015)
86. J. Meyer, A. Rufer, A DC hybrid circuit breaker with ultra-fast contact opening and integrated gate-commutated thyristors (IGCTs). IEEE Trans. Power Del. **30**(2), 646–651 (2006)
87. L. Novello, F. Baldo, A. Ferro, Development and testing of a 10-kA hybrid mechanical–static DC circuit breaker. IEEE Trans. Appl. Supercond. **21**(6), 3621–3627 (2011)
88. R. Schmerda, R. Cuzner, R. Clasrk, D. Nowak, S. Bunzel, Shipboard solid-state protection: overview and applications. IEEE Elect. Mag. **1**(1), 32–39 (2013)

89. Z.J. Shen, G. Sabui, Z. Miao, Z. Shuai, Wide-bandgap solid-state circuit breakers for DC power systems: device and circuit considerations. IEEE Trans. Electron Dev. **62**(2), 294–300 (2015)
90. A. Mokhberdoran, N. Carvalhoa, H. Silva, A. Leitea, Carrapatosoa, design and implementation of fast current releasing DC circuit breaker. Electr. Power Syst. Res. **151**, 218–232 (2017)
91. S.Y.R. Hui, C.K. Lee, F. Wu, Electric springs—A new smart grid technology. IEEE Trans. Smart Grid **3**(3), 1552–1561 (2012)
92. G. Ma, G. Xu, Y. Chen, R. Ju, Voltage stability control method of electric springs based on adaptive PI Controller. Int. J. Elec. Power Energy Syst. **95**, 202–212 (2018)
93. J. Liao, N. Zhou, Y. Huang, Q. Wang, Decoupling control for DC electric spring-based unbalanced voltage suppression in a bipolar DC distribution system. IEEE Trans. Ind. Electron. **68**(4), 3239–3250 (2021)
94. Z. Liu, I. Cvetkovic, Z. Shen, D. Boroyevich, R. Burgos, Imbalance mechanism and balancing control of DC voltages in a transformerless series injector based on paralleled H-bridge converters for AC impedance measurement. IEEE Trans. Power Electron. **34**(8), 8175–8189 (2019)
95. H. Bai, X. Wang, P.C. Loh, F. Blaabjerg, Passivity enhancement of grid-tied converters by series LC-filtered active damper. IEEE Trans. Ind. Electron. **64**(1), 369–380 (2017)
96. K. Wang, X. Zhang, H.S. Chung, Solid-state single-port series damping device for power converters in DC microgrid systems. IEEE Trans. Power Electron. **34**(1), 192–202 (2019)
97. H. Wang, H.S. Chung, W. Liu, Use of a series voltage compensator for reduction of the DC-link capacitance in a capacitor-supported system. IEEE Trans. Power Electron. **29**(3), 1163–1175 (2014)
98. Z. Li, W. Li, T. Pan, An optimized compensation strategy of DVR for micro-grid voltage sag. Pro. Control Mod. Power Sys. **1**(1), 78–85 (2016)
99. M. Shahabadini, H. Iman-Eini, Improving the performance of a cascaded H-bridge-based interline dynamic voltage restorer. IEEE Trans. Power Del. **31**(3), 1160–1165 (2016)
100. H. Hafezi, R. Faranda, Dynamic voltage conditioner, a new concept for smart low-voltage distribution system. IEEE Trans. Power Electron. **33**(9), 7582–7590 (2017)
101. C. Meyer, R.W. Doncker, Y.W. Li, F. Blaabjerg, Optimized control strategy for a medium-voltage DVR—Theoretical investigations and experimental results. IEEE Trans. Power Electron. **23**(6), 2746–2754 (2008)
102. G.A.A. Carlos, C.B. Jacobina, E.C. Santos, Investigation on dynamic voltage restorers with two DC links and series converters for three-phase four-wire systems. IEEE Trans. Ind. Appl. **52**(2), 1608–1621 (2016)
103. P. Li, L. Xie, J. Han, S. Pang, P. Li, New decentralized control scheme for a dynamic voltage restorer based on the elliptical trajectory compensation. IEEE Trans. Ind. Electron. **64**(8), 6484–6495 (2017)
104. R.K. Agarwal, I. Hussain, B. Singh, Implementation of LLMF control algorithm for three-phase grid-tied SPV-DSTATCOM system. IEEE Trans. Ind. Electron. **64**(9), 7414–7424 (2017)
105. B. Liu, L. Wang, D. Song, et al., Input current ripple and grid current harmonics restraint approach for single-phase inverter under battery input condition in residential photovoltaic/battery systems. IEEE Trans. Sustain. Energ. **9**(4), 1957–1968 (2018)
106. M.J. Newman, D.G. Holmes, J.G. Nielsen, F. Blaabjerg, A dynamic voltage restorer (DVR) with selective harmonic compensation at medium voltage level. IEEE Trans. Ind. Appl. **41**(6), 1744–1754 (2005)
107. M. Pradhan, M.K. Mishra, Dual P-Q theory based energy optimized dynamic voltage restorer for power quality improvement in distribution system. IEEE Trans. Ind. Electron. **66**(4), 2946–2955 (2019)
108. P. Cheng, C. Huang, C. Pan, S. Bhattacharya, Design and implementation of a series voltage sag compensator under practical utility conditions. IEEE Trans. Ind. Electron. **39**(3), 844–853 (2003)

109. L.A. Moran, I. Pastorini, J. Dixon, R. Wallace, A fault protection scheme for series active power filters. IEEE Trans. Power Elect. **14**(5), 928–938 (1999)
110. M.J. Newman, D.G. Holmes, An integrated approach for the protection of series injection inverter. IEEE Trans. Ind. Electron. **38**(3), 679–687 (2002)
111. Axente, M. Basu, M.F. Conlon, Gaughan, Protection of unified power quality conditioner against the load side short circuits. IET Power Electron. **3**(4), 542–551 (2010)
112. S. Alaraifi, A. Moawwad, M.S.E. Moursi, V. Khadkikar, Voltage booster schemes for fault ride-through enhancement of variable speed wind turbines. IEEE Trans. Sustain. Energy **4**(4), 1071–1081 (2013)
113. B. Yan, D. Jiang, Z. Wu, Y. Zang, D. Gan, Design of unified power flow controller with fault current limiting. Autom. Electric Power Sys. Press **36**(4), 69–73 (2012)
114. Chen, D. Jiang, Y. Fan, K. Chen, Y. Guo, W. Lv, Design of a novel bridge-type FCL and its application in UPFC, in *Power and Energy Society General Meeting* (2015), pp. 1–5
115. Chen, D. Jiang, Z. Lu, Z. Wu, A new proposal for solid state fault current limiter and its control strategies, in *Power Engineering Society General Meeting* (2004), pp. 1468–1473
116. S. Xu, Q. Du, X. Zhang, H. Ma, Y. Wang, Current limiting analysis and parameter design of a novel UPFC with fault current limiter. Power Syst. Technol. Press **41**(2), 558–565 (2017)
117. W. Lv, C. Shen, D. Jiang, F. Gui, Y. Fan, Z. Wu, UPFC with fault current limiting and capacitor voltage limiting. Zhejiang University XueBao **48**(5), 877–881 (2014)
118. S.S. Choi, T.X. Wang, D.M. Vilathgamuwa, A series compensator with fault current limiting function. IEEE Trans. Power Del. **20**(3), 2248–2256 (2005)
119. M.R. Banaei, A. Nabavandi, S.H. Hosseini, Investigation of multi-functional DVR to improve power quality characteristic, in *Electrical Engineering/Electronics, Computer, Telecommunications and Information Technology (ECTI-CON)* (2009), pp. 144–149
120. D.M. Vilathgamuwa, P.C. Loh., Y. Li, Protection of microgrids during utility voltage sags. IEEE Trans. Ind. Electron. **53**(5), 1427–1436 (2006)
121. M. Zafari, E. Gholipour, S.M. Madani, Active fault current limiter and comparison of its performance with SFCL, in *Power Signals Control and Computations (EPSCICON)* (2015), pp. 1–6
122. Y. Li, D.M. Vilathgamuwa, P. Loh, F. Blaabjerg, A dual-functional medium voltage level DVR to limit downstream fault currents. IEEE Trans. Power Electron. **22**(4), 1330–1340 (2007)
123. F.M. Mahdianpoor, R.A. Hooshmand, M. Ataei, A new approach to multifunctional dynamic voltage restorer implementation for emergency control in distribution systems. IEEE Trans. Power Del. **26**(2), 882–890 (2011)
124. W. Zhou, J. Le, J. Yang, X Du., K. Liu, Control strategy of MV dynamic voltage restorer as virtual impedance. Power Syst. Technol. **40**(8), 2525–2532, (2016)
125. H. Nourmohamadi, M. Sabahi, P.T. Balsara, E. Babaei, S.H. Hossein, A. Fakhim-Babaei, New concept for fault current limiter with voltage restoration capability. IEEE Trans. Ind. Electron. **16**(12), 10001–10010 (2020)

Part I
Optimization Operation Strategy for Voltage Quality Controller

With the main objective of mitigating the phase jump in the load-side voltage while improving the overall sag compensation time, an efficient strategy to improve the voltage quality of sensitive loads with the optimal utilization of a VQC is proposed firstly. Then, the output power quality of the cascaded H-bridge VQC can be guaranteed by maximizing the output voltage level number as elaborated in Chap. 3.

Chapter 2
The Mitigation Strategy of Voltage Sags and Phase Jumps for Voltage Quality Controllers

This chapter introduces an efficient strategy to improve the voltage quality of sensitive loads with the optimal utilization of a VQC, with the main objective of mitigating the phase jump in the load-side voltage while improving the overall sag compensation time.

2.1 Introduction

The rapid development of renewable energy has increased the complexity of the power system and may lead to a decrease in the quality of electricity in the future distribution systems [1, 2]. To increase voltage reliability, one effective technique is the employment of a custom-made series active compensator, such as the VQC [3, 4] illustrated in Fig. 2.1.

For the stable operation of the power system and the improvement of power quality, it can be seen from literature [5–10] that researchers have made a lot of efforts on the VQC control strategy. Early research mainly focused on two aspects: one is in-phase compensation, which reduces the rated value of VQC injection voltage [11], and the other is to solve phase jump to achieve preset compensation, which ensures that the load-side voltage can recover to its preset value [10]. These approaches typically demand higher energy storage capacities due to their larger active power outputs during compensation. To mitigate deep sags with extended durations while minimizing energy storage requirements, a new compensation strategy has been proposed to achieve reactive power compensation and keep active power at the lowest level [12, 13], referred to as an energy-optimized strategy [14–16]. It is worth noting that due to the difference in the X/R ratio between the source impedance and the feeder impedance, voltage sag events usually cause phase jumps [17]. Under the action of energy optimization compensation strategies, this phenomenon may be further exacerbated, leading to significant phase jumps in the load-side voltage [18].

F. Jiang et al., *Advanced Voltage Quality Controller for New Power Distribution Systems*, https://doi.org/10.1007/978-981-97-9861-2_2

Fig. 2.1 Typical configuration of VQC-based system in single-phase

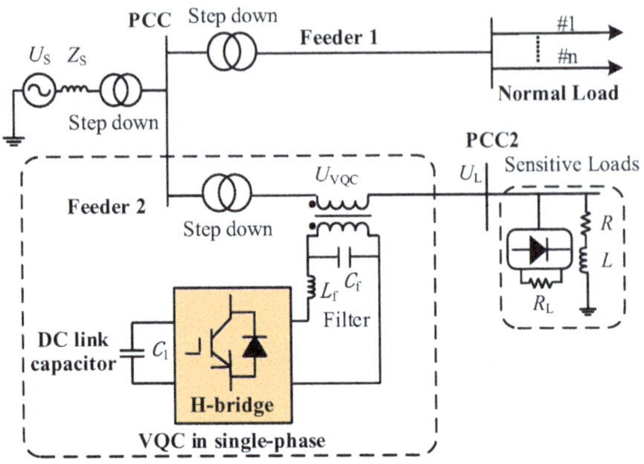

Given the occurrence of transient currents, particularly in the case of most loads [19], it is imperative to address large phase jumps and avoid them [14]. Therefore, there is growing interest in optimal control methods that extend overall sag compensation time while rectifying phase jumps.

Certain researchers [18] allowed the further utilization of the DC link capacitor energy, which reduced the needed injection voltage of the VQC as the DC link voltage reaches a certain limit, and concurrently mitigated the transient distortions at the load side and prolonged the compensation time. However, the energy stored in the DC link capacitor continues to be consumed throughout the compensation stage, and the compensation time enhancement is thus limited [14]. An idea of cross-phase voltage injection is applied in an interphase AC–AC topology to provide increased capability for pre-sag compensation and phase jump mitigation [20]. Considering the proposed topology with a high coupling coefficient between phases, the methods to embody the superiority of this strategy in single-phase systems need to be studied.

An enhanced strategy using a smooth transition from pre-sag compensation to the energy-optimized scheme is proposed to mitigate the phase jump caused by sag occurrence; the VQC sag support time is significantly increased compared with that of existing methods in references [13, 14]. Next, in reference [21], a new generalized philosophy that controls the phase angle and magnitude of the injected voltage is proposed, which accomplishes sag correction based on the proper application of elliptical compensation. The method makes a VQC go through a smooth transition from standby to voltage compensation, and maintains the load voltage at a desired level without a noticeable phase angle or magnitude. It must be noted that the phase jump in the load side after the voltage fluctuation removal is bound to exist because of sag issues and energy-optimized compensation strategies, but not much attention has been paid to this issue in previous studies. Consequently, the smooth exit of a VQC from the power grid after fault elimination needs further research. To perfectly satisfy a modern power system, a proper method should cover long-term as well as short-term sag events and mitigate phase jumps during the entire process of voltage sag cases.

Moreover, to guarantee device stability, in reference [22], a new updated procedure for the load reference voltage is suggested that regulates the initial operating point of a VQC if the system goes outside its limits. Similarly, the limitation case and updated procedure are considered in the recovery stage [23]. Overall, operating boundaries may exist in both stages and cannot be considered separately to avoid the phase jump caused by the mode switching, which is the key issue that is not taken into account in references [24, 25].

Therefore, a comprehensive scheme to mitigate voltage sag and phase jump is proposed in this chapter, which features the following superior characteristics:

(1) A flexible recovery strategy is vital for a VQC to avoid the phase jump in the load-side voltage after sag removal. The linkage between the voltage compensation and recovery stage needs to be well designed, and the initial operating point of the latter is adjusted according to the final operating point of the former to realize the phase jump mitigation in both stages.
(2) To guarantee device stability, the prior calculation of the maximum amplitude of the system output in both stages is executed. An updated procedure for the injected reference voltage is suggested to prevent the system from going outside its operation limits, and a solution to resolve the overmodulation caused by the DC link voltage reduction is considered as well.

2.2 Traditional Operation Method

Figure 2.1 presents a schematic diagram of the configuration of single-phase VQC, with similar settings for other phases. Generally speaking, the energy required for voltage compensation in VQC can be obtained through the DC link or extracted from the energy storage system connected to the DC link, which is supported by references [1, 2]. This chapter specifically employs a capacitor-supported VQC, and a series transformer is utilized to achieve electrical isolation between the VQC and the grid side, as detailed in references [4, 26]. Over the years, various strategies have been employed by VQC systems for voltage compensation, as documented in reference [1]. As the most commonly used methods, there are in-phase compensation, reactive power compensation, pre-sag compensation, and minimum active power compensation [10], particularly in-phase compensation and pre-sag compensation, have demonstrated the potential to significantly enhance the ride-through capability of a VQC [14]. This section provides an overview of the operational principles of the voltage compensation strategy, as well as the traditional method for mitigating phase jumps, offering a comprehensive understanding of the VQC's functionality and its capacity for addressing voltage fluctuations in power systems.

Case 1: Earlier works [18] have shown that the sag can be compensated using only a reactive power injection if the sag depth k_{sag} is within certain power factor-dependent limits. The relation is as follows:

$$k_{\text{sag}} = \left(U_{\text{Lref}} - U_S'\right) / U_{\text{Lref}}, \tag{2.1}$$

and k_{sag} is defined as

$$k_{\text{sag}} \leq 1 - \cos(\theta_L), \tag{2.2}$$

where θ_L is the power factor angle of the load side, $K_{\text{sag}} \leq 1 - \cos(\theta_L)\ U_{\text{Lref}}$, is the reference value of load voltage, U_{S} and U_{S}' represent the rated and sagged grid voltages, respectively.

The phasor representation of VQC operation is shown in Fig. 2.2a, the grid voltage sags accompanying with a phase jump (σ), the injected voltage of VQC (U_{VQCf}) is perpendicular to load current after sag (I_L'), α_r is the phase jump due to the energy-optimized compensation, U_{presag} is the grid voltage before the sag occurs.

The RMS value and the angle of the injected voltage are given by the following:

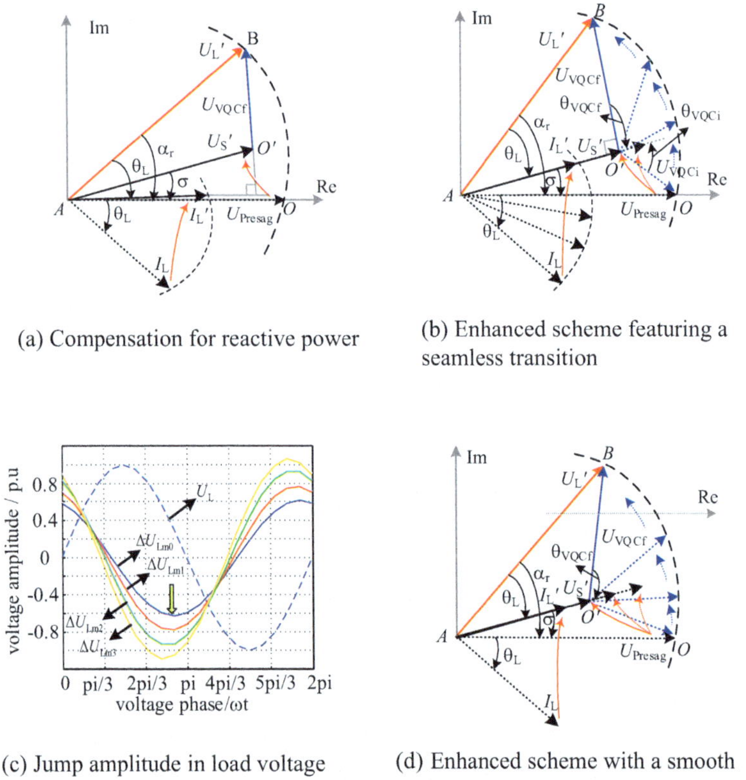

(a) Compensation for reactive power

(b) Enhanced scheme featuring a seamless transition

(c) Jump amplitude in load voltage

(d) Enhanced scheme with a smooth transition while $k_{sag} > 1 - cos(q_L)$

Fig. 2.2 Phasor diagrams representing the voltage compensation stage

$$U_{\mathrm{VQCf}} = U_L' \sin\theta_L - \sqrt{(U_S')^2 - (U_L' \cos\theta_L)^2} \tag{2.3}$$

$$\theta_{\mathrm{VQCf}} = \pi - \arccos\left(\frac{(U_S')^2 + (U_{\mathrm{VQCf}})^2 - (U_L')^2}{2U_S' U_{\mathrm{VQCf}}}\right), \tag{2.4}$$

Respectively, the phasors U_L and U_L' denote the load voltage prior to and subsequent to the sag, as referenced in Ref. [10]. In this case, the capacitor-supported VQC is able to cover the short- and longer -term voltage variations.

In the scenario illustrated in Fig. 2.2b, the limiting case involves quadrature injection with $k_{\mathrm{sag}} = 1 - \cos(\theta_L)$, resulting in $\theta_{\mathrm{VQCf}} = \pi/2$. In this situation, the VQC plays a crucial role in effectively supporting reactive power under full load conditions, ensuring that the power grid operates at a unit power factor. In order to gain a deeper understanding of the relationship between phase jumps and voltage amplitude flicker, we utilize the feature of ΔU_{Lm} to quantify deviations in specific situations, in order to better understand the dynamic response and stability of the power grid. The formula is expressed as follows:

$$\Delta U_{\mathrm{Lm}} = \sqrt{2} U_{\mathrm{Lref}}(\sin(\omega t + \theta_L + \sigma) - \sin(\omega t)), \tag{2.5}$$

where ω represents the angular frequency. Figure 2.2c shows the values of ΔU_{Lm0} to ΔU_{Lm3} with a variation in σ (0°, 10°, 20°, and 30°). The graph demonstrates that the amplitude of the load voltage jump is directly proportional to the phase jump angle. Additionally, the implementation of an energy-optimized strategy plays a crucial role in determining the trend of the amplitude deviation. In Fig. 2.3, a voltage fluctuation occurs under $k_{\mathrm{sag}} = 0.2$ with a phase jump of 10 degrees between 0.42 and 0.50 s. Although the voltage amplitude of U_L is effectively maintained using the traditional energy-optimized strategy, it is noted that the phase jump during the initial stage of voltage compensation cannot be entirely avoided.

The goal of the enhancement scheme proposed in reference [14] is to solve the voltage phase jump problem by promoting a seamless transition from pre depression compensation to energy optimization compensation. The implementation of this scheme can be observed in Fig. 2.2b, where the blue curve trajectory shows the effect of this transition. Using the relation among U_{presag}, U_{VQCi} and U_S', the RMS value and the angle of the injected voltage are given in the following:

$$U_{\mathrm{VQCi}} = \sqrt{(U_{\mathrm{presag}})^2 + (U_S')^2 - 2U_{\mathrm{presag}} U_S' \cos(\sigma)}, \tag{2.6}$$

$$\theta_{\mathrm{VQCi}} = -\left(\pi - \arccos\left(\frac{(U_{\mathrm{VQCi}})^2 + (U_S')^2 - (U_{\mathrm{presag}})^2}{2U_S' U_{\mathrm{VQCi}}}\right)\right). \tag{2.7}$$

The variation tendency of the voltage compensation vector is depicted in Fig. 2.2b, which starts from O′O, goes through arc OB and terminates at O′B. A transition ramp is defined from the initial operating point to the final one to ensure the smooth

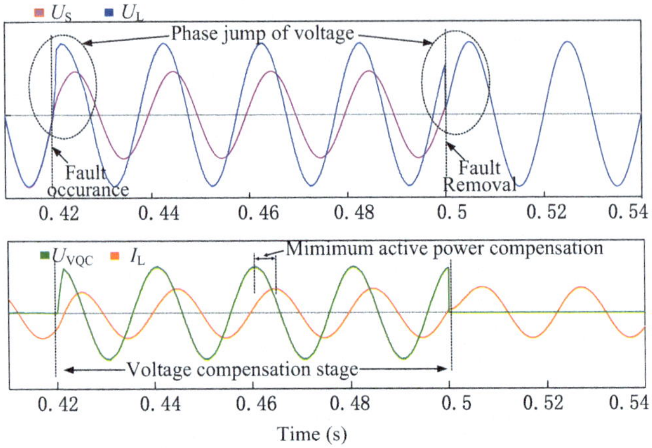

Fig. 2.3 Performance of VQC with traditional energy-optimized strategies

changeover is completed [14]:

$$\theta_{\text{trans1}} = \theta_{\text{VQCi}} + \frac{\theta_{\text{VQCf}} - \theta_{\text{VQCi}}}{\Delta t_1}(t). \tag{2.8}$$

The slope of the transition curve is determined by Δt_1. And the faster the transition completes, the longer the VQC compensation time is, as the energy absorbed from the DC link is reduced.

Case 2: When the voltage sag occurs with $k_{\text{sag}} > 1 - \cos(\theta_L)$, sag correction using only reactive power is impossible. To achieve compensation with a minimum of active power, U_S' must be made to lie along the direction of I_L', and U_{VQCf} is no longer perpendicular to I_L', as shown in Fig. 2.2d. Using the relation among U_{presag}, U_{VQCf}, and U_S', the RMS value of the injected voltage can be found as follows:

$$U_{\text{VQCf}} = \sqrt{\left(U_S'\right)^2 + \left(U_L'\right)^2 - 2\left(U_S'\right)\left(U_L'\right)\cos(\theta_L)}, \tag{2.9}$$

where the per unit values of U_S' and U_L' after voltage compensation is equal to $1 - k_{\text{sag}}$ and 1 p.u, respectively.

Accordingly, substituting (2.9) into (2.4), the angle of U_{VQCf} can be calculated. Then, a smooth changeover can be achieved, similar to Case 1. Note that the certain energy is absorbed from the DC link of VQC during the transition stage and the steady operating stage, especially the latter.

2.3 Proposed Operation Method

This chapter introduces a holistic scheme aimed at alleviating voltage sags and phase jumps through the above strategy. The scheme encompasses a well-thought-out connection between them. Additionally, updated procedures for the injected reference voltage are proposed to ensure that the system remains within its operational limits. Subsequent subsections delve into the details of these strategies and their implementation for effective mitigation of voltage disturbances.

2.3.1 Proposed Operation Principle

After using the smooth transition, U_L and I_L with a negligible phase jump at the initial stage of the voltage compensation are shown in Fig. 2.4. However, a phase jump on U_L after the fluctuation removal may still exist, as shown in Fig. 2.3, which has been seldom studied. Overall, the large phase jump may occur two times during a whole compensation stage. A comprehensive scheme for a VQC should ensure the load-side voltage quality with the phase jump mitigation in the compensation stage, but also include a recovery strategy that makes the VQC exit smoothly to standby mode. So, another smooth transition and the linkage between voltage compensation and recovery stage are designed in this chapter.

The system voltage U_S' is back to U_{presag} after the sag case removal [23]. Based on the transition in voltage compensation stage, it is possible to instantaneous injection of voltage U_{ri} can effectively avoid phase jumps that occur after fault elimination, as shown in Fig. 2.5. In the initial state of the recovery phase, the injected voltage follows

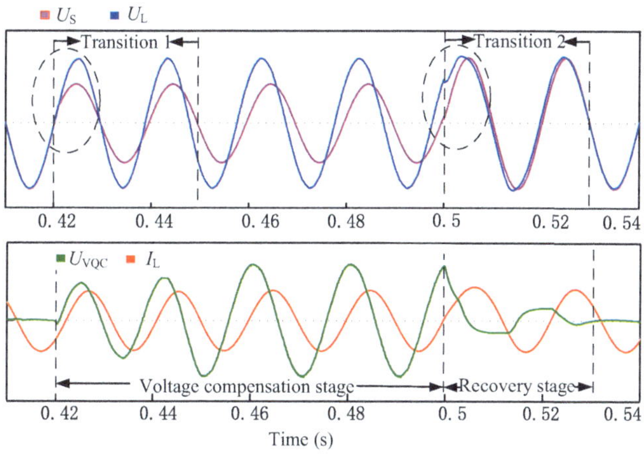

Fig. 2.4 Performance of VQC with enhanced voltage compensation scheme and proposed recovery strategy

Fig. 2.5 Phasor diagrams for the initial state in the VQC recovery stage

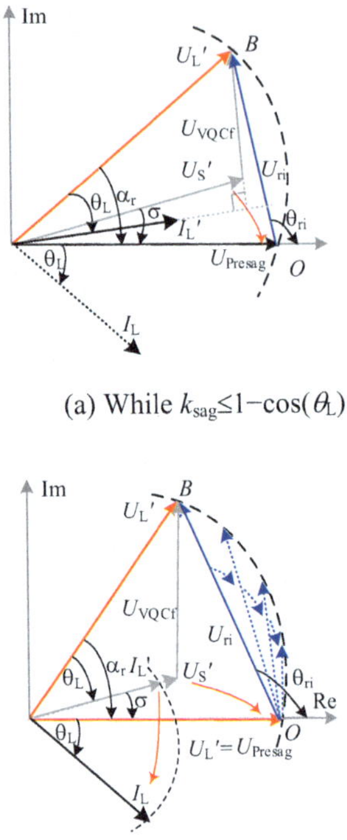

(a) While $k_{\text{sag}} \le 1-\cos(\theta_L)$

(b) While $k_{\text{sag}} > 1-\cos(\theta_L)$

a specific relationship, which helps the system recover to a stable operating state. The correct application of injected voltage can provide important control measures in power systems to ensure stable voltage and phase conditions. Specifically, as follows:

$$\vec{U}_{\text{presag}} + \vec{U}_{\text{ri}} = \vec{U}'_{L}. \tag{2.10}$$

As shown in Fig. 2.5a and b, α_r is given by:

$$\alpha_r = \begin{cases} \theta_L + \sigma & \text{if } k_{\text{sag}} > 1 - \cos(\theta_L) \\ \theta_L + \theta_{\text{VQCf}} - \frac{\pi}{2} + \sigma & \text{if } k_{\text{sag}} \le 1 - \cos(\theta_L) \end{cases}. \tag{2.11}$$

Accordingly, θ_{ri} and U_{ri} can be calculated by (2.11)

$$\theta_{\text{ri}} = \pi - \frac{\pi - \alpha_r}{2}, \tag{2.12}$$

$$U_{\text{ri}} = \sqrt{(U'_L)^2 + (U_{\text{presag}})^2 - 2(U'_L)(U_{\text{presag}})\cos(\alpha_r)}, \tag{2.13}$$

where the per unit value of U_{presag} is equal to 1 p.u. The unit value of U_{presag} is 1p. u., we take the smooth switching under the condition of $k_{sag} > 1 - \cos(\theta_L)$ sag as an example to analyze the trend of voltage compensation vector variation. This process is shown in Fig. 2.5b. The switching process starts from the starting point OB, passes through the arc BO, and finally reaches point O. After the switch is completed, the angle of the final operating point needs to be adjusted according to the following specific conditions to ensure that the stability and performance of the system meet the expected requirements.

$$\theta_{rf} = \frac{\pi}{2}. \tag{2.14}$$

Then, to implement the smooth exit operation of the VQC, the variation tendency of θ_{VQC} is depicted in Fig. 2.6a with a blue line during Δt_2, and the transition ramp is defined as

$$\theta_{trans\,2} = \theta_{rf} + \frac{\theta_{rf} - \theta_{ri}}{\Delta t_2}(t). \tag{2.15}$$

Figure 2.5b shows the result of a smooth conversion, i.e., the position conversion of U_L' to U_{presag}

A flexible linkage between the voltage compensation and recovery stage is proposed in this chapter that can perfectly react to short-term sag cases while mitigating the phase jump at the protected load (Case A, as shown in Fig. 2.6 from t_0 to t_2). The performance of a VQC after fault removal with the proposed strategy, regardless of the operation limits of the VQC system, is shown in Fig. 2.4. The large voltage flicker is mitigated in comparison with the effect shown in Fig. 2.3.

Furthermore, for a long sag event with a deeper k_{sag} (case B shown in Fig. 2.6 from t_0 to t_5, exceeding the limit given in (2.1)), note that the energy stored in the DC link capacitor is reduced due to a certain nonzero active power injected by the VQC, and consequently, the DC link voltage decreases (gradually). Though the rate of decline can be reduced by increasing the capacitance capacity, it is still unavoidable. So it is better to have an adjusted strategy to prevent overmodulation, which will further stabilize the operation of the VQC. To maintain the rated load voltage and prevent the controller from going into overmodulation, the following relation should be satisfied:

$$\sqrt{2}U_{VQC} \leq m_{max} U_{dcmin} = \sqrt{2}U_{VQCmax}, \tag{2.16}$$

where U_{dcmin} is the limit setting value of the DC link voltage. By the derivation of (2.4), the value of $d\cos(\theta_{VQCf})/dU_{VQCf}$ is given in (2.17):

$$\frac{d\cos(\theta_{VQCf})}{dU_{VQCf}} = \frac{(U_L')^2 - (U_S')^2}{2U_S'(U_{dc}m_{max})^2} + \frac{m_{max}(1 - U_S'(U_{dc}m_{max}))}{2(U_S')^2 U_{dc}m_{max}}. \tag{2.17}$$

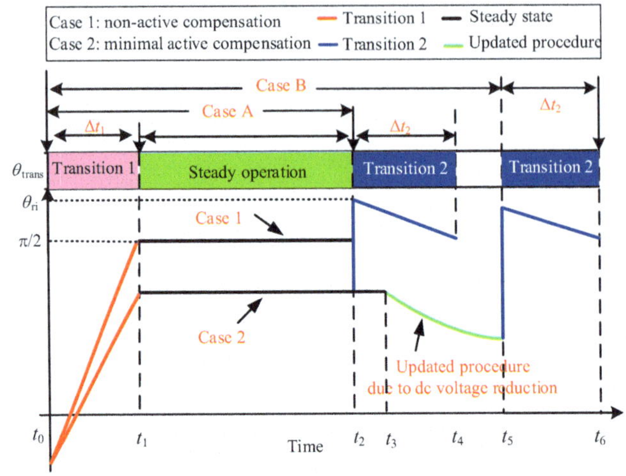

Fig. 2.6 Injection angle of VQC with a smooth transition in the voltage compensation and recovery stage

Fig. 2.7 Phasor diagrams of operation correction due to DC link voltage reduction

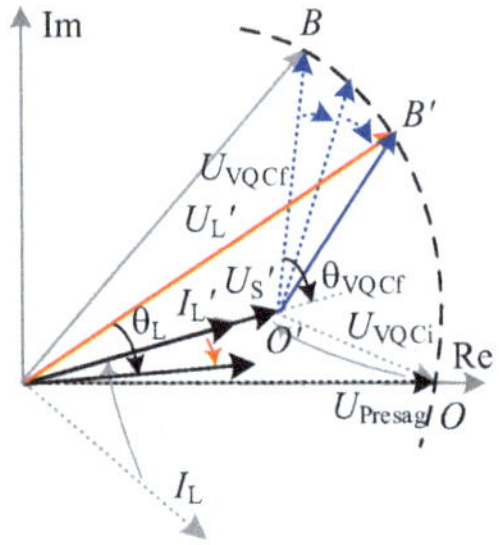

(a) Voltage compensation stage

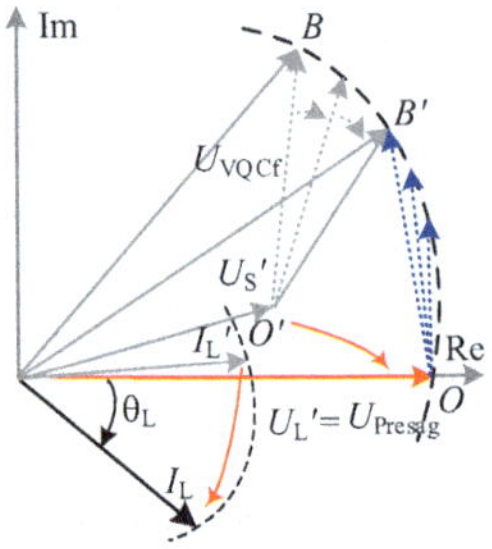

(b) Recovery stage

From (2.17), we can see that the lower the value of U_{dc}, the greater is the changeover slope of $\cos(\theta_{VQCf})$, which represents a decreasing tendency of θ_{VQCf}, as can be noticed from Fig. 2.6 with the green line from t_3 to t_5. Hence, during the voltage compensation stage, the injected voltage U_{VQCf} must be smoothly changed, which can be realized by unblocking θ_{VQCf}, as depicted in Fig. 2.7a, where the voltage compensation vector starts from O′B, goes through arc BO and terminates at point O′B′. Then, the needed amplitude of the injected voltage is reduced to make the DC link discharge even more and extend the VQC compensation time. Accordingly, the recovery process should also be adjusted, and the smooth transition is initiated as the changeover from t_5 to t_6 using (2.15), as shown in Fig. 2.7b. The overall calculation and implementation of the updated procedure are discussed in the following subsections. The design principle of capacitance volume and the method to recover DC link voltage will not be covered by this work, because several elaboration and solutions are just presented, as in references [7, 14, 22, 23], and they can be added to the proposed strategy.

2.3.2 *VQC Overrun Operation Stage*

In reality, the overmodulation operation of the VQC is probably caused by the DC link voltage reduction, and the design limit being exceeded at the initial operating point in both stages is one of the reasons as well. From Figs. 2.2 and 2.5, we can see that the proposed strategy for the VQC depends on the k_{sag}, σ and θ_L, and has the possible operation limits on U_{VQCf} and U_{ri}. To compare the two parameters as indicated in (2.9) and (2.13), the graph of U_{VQCf} and U_{ri} with the changes of θ_L, k_{sag} and δ is illustrated in Fig. 2.8a in p. u, which gives values of $\theta_L = (0, 60°)$, $\sigma = (0, 30°)$, and $k_{sag} = (0.05, 0.5)$. It can be seen that the larger θ_L or δ is, the more likely U_{ri} is to exceed the operating limits.

Therefore, it is necessary to ensure that the compensation limit of VQC conforms to the formula 2.2–2.18:

$$\max(U_{VQC}) = \max(U_{VQCf}, U_{ri}) \leq U_{VQCmax}. \tag{2.18}$$

The dotted circle represents the voltage operation limits with radius U_{VQCmax} at points O and O′. Once U_{ri} or U_{VQCf} exceeds the operation limits, U_{VQCmax} is taken as the value of U_{ri} or U_{VQCf}, as shown in Fig. 2.8b and c, respectively. The initial operating state of the voltage compensation and recovery stage should be regulated, and α' can be given by

$$\alpha' = \begin{cases} \arccos\left(\frac{2U_{Lref} - (U_{VQC,max})^2}{2U_{Lref}}\right) & \text{if } \max(U_{VQC}) = U_{ri} \\ \delta + \arccos\left(\frac{(U_S')^2 + (U_L')^2 - (U_{VQC,max})^2}{2U_S'U_L'}\right) & \text{if } \max(U_{VQC}) = U_{VQCf} \end{cases}. \tag{2.19}$$

Fig. 2.8 Updated procedure

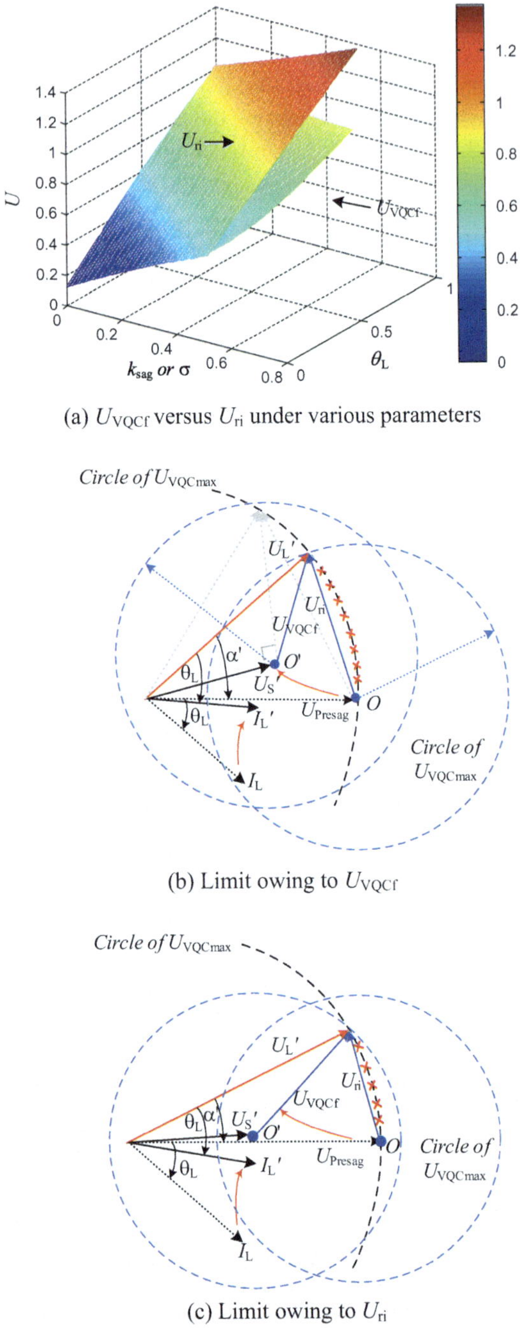

(a) U_{VQCf} versus U_{ri} under various parameters

(b) Limit owing to U_{VQCf}

(c) Limit owing to U_{ri}

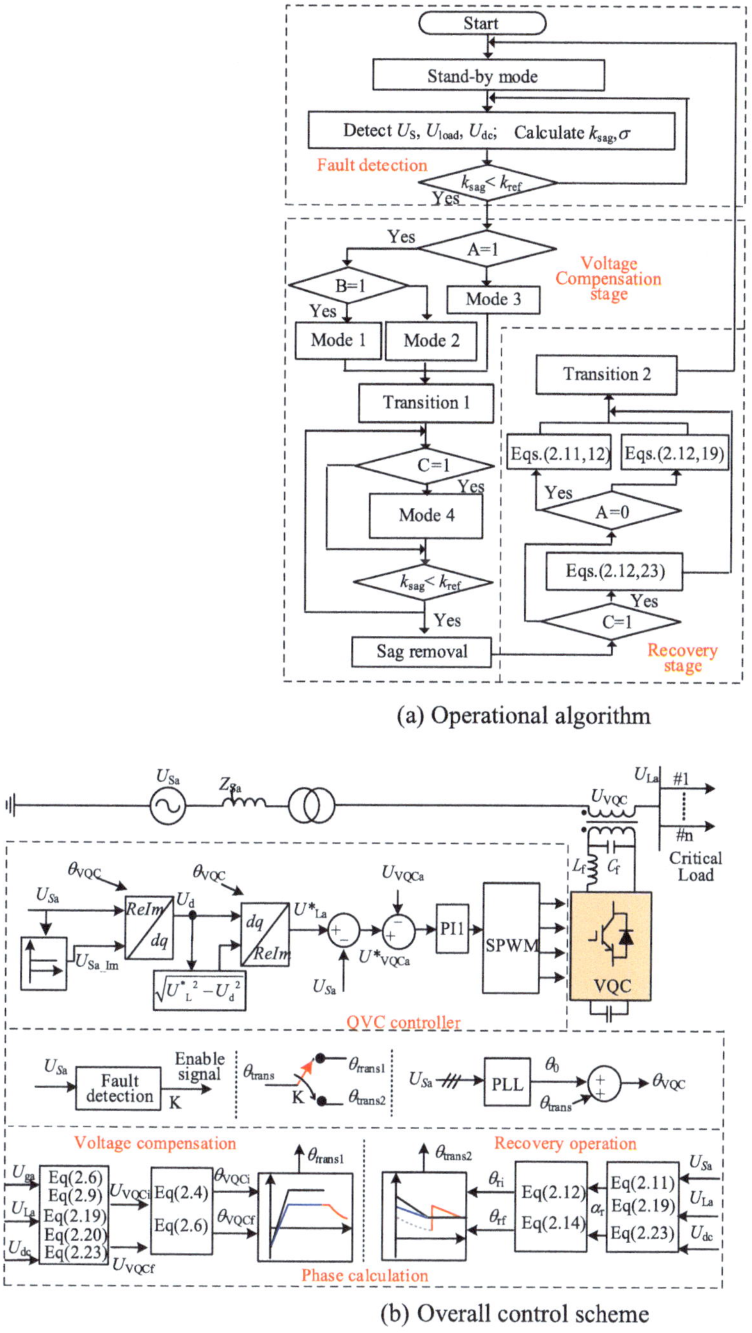

Fig. 2.9 Block diagram of the adaptive compensation scheme

Then, the RMS value of the injected voltage should be readjusted because of the regulation as

$$U_{\mathrm{VQCf}} = \begin{cases} \sqrt{(U_L')^2 + (U_S')^2 - 2U_L'U_S'cos(\alpha' - \sigma)} & \text{if}\max(U_{\mathrm{VQC}}) = U_{\mathrm{ri}} \\ U_{\mathrm{VQC\,max}} & \text{if}\max(U_{\mathrm{VQC}}) = U_{\mathrm{VQCf}} \end{cases}. \tag{2.20}$$

Accordingly, substituting (2.20) into (2.4), θ_{VQCf} can be recalculated. Then, the updated procedure caused by the DC link voltage reduction should be considered as well. Firstly, U_{dcmin} is reconfirmed and given as

$$U_{\mathrm{dcmin}} = \begin{cases} \frac{\sqrt{2}U_{\mathrm{VQC\,max}}}{m_{\max}} & \text{if}\max(U_{\mathrm{VQC}}) \geq U_{\mathrm{VQC\,max}} \\ \sqrt{2}\max(U_{\mathrm{VQC}}) & \text{if}\max(U_{\mathrm{VQC}}) < U_{\mathrm{VQC\,max}} \end{cases}, \tag{2.21}$$

Once the value of U_{dc} is less than U_{dcmin}, the following relation should be satisfied at all times:

$$\sqrt{2}\max\left(U_{\mathrm{VQC}}\right) = \sqrt{2}U_{\mathrm{VQCmax}} \leq m_{\max}U_{\mathrm{dc}}, \tag{2.22}$$

Then, using (2.19), the updated procedure of $\alpha'$is confirmed as

$$\alpha' = \min(\alpha'(U_{\mathrm{VQCf}}), \alpha'(U_{\mathrm{ri}})). \tag{2.23}$$

Hence, substituting (2.20) and (2.23) into (2.4), we can obtain the updated value of θ_{VQCf} in real time. Finally, taking the value of α' at the instant of the sag removal as α_r, the elastic recovery is injected voltage reference is suggested to prevent the system going outside its operation limits, which can further guarantee the stability of the device.

2.4 Overall Operation Logic and Control Scheme

The VQC may run in various modes according to the states of the grid and DC link. In this section, the operation procedure of the transition from pre-sag compensation to energy-optimized strategies, the transition to make the VQC smooth exit to standby mode is briefly expounded. Further, the operation logic and overall control scheme are explained in the following subsections as well.

2.4.1 Operation Logic Analysis

Owing to the randomness of voltage sag issue, U_S and U_L are measured in real time. Under normal conditions, the VQC works in standby mode. The voltage compensation stage is initiated while $k_{sag} > k_{ref} = 0.05$. Actually, in this chapter, as soon as the sag is detected, the first step is to implement the pre-sag compensation strategy to restore the load voltage and phase to the pre-sag values. Conventionally, for the pre-sag strategy, the PLL is frozen by using a phase freezer unit, which fixes(freezes) the pre-sag phase angle of the supply voltage, to be used as the reference phase angle of the load voltage, as described in references [27–29]. Then, the controller initiates the transition from pre-sag compensation to energy-optimized compensation after the phase jump is calculated [30]. So, the desired response of VQC PLL system is quite different from other applications [26]. In this chapter, the Deri-PLL is adopted [31], and in the case of grid voltage distortion, the method presented in references [31–36] may be a better choice. Overall, the expected compensation performance of VQC in distorted grid conditions can be ensured by combining the above-mentioned PLL or completed by using (2.12) and (2.15). Overall, the purpose of this work is obvious while compared with the conventional algorithms as explained in 2.2. On the one hand, the proposed strategy analyses and solves the phase jump caused by the deactivation of VQC once the sag disappears, which is the further improvement of the energy-optimized compensation strategy. On the other hand, the prior calculation of the maximum output power of VQC in both stages is executed, and an updated procedure for the fundamental extraction methods with the proposed strategy.

The operation mode and the initial operating point are determined based on the values of k_{sag}, $\cos(\theta_L)$ and $U_{VQC\ max}$. The prior calculation of the VQC's maximum output in both stages and the adjustment of the initial operating point avoid the occurrence of the overmodulation case. Furthermore, once the DC link voltage drops below a given value $U_{dc\ min}$, the injected voltage angle during the voltage compensation stage is regulated because an iterative control loop is activated, thus preventing overmodulation from occurring again. Finally, as for the procedure of transition to make VQC system smooth exit to standby mode, which is similar to the procedure of controlling the transition from standby to compensation mode, the first step is to restore the load voltage and phase angle equal to the pre-recovery values by using the value of α_r at the instant of sag removal. Then, the controller enables another transition to make the VQC smooth exit to standby mode. The smooth transition is adopted two times to ensure that the VQC protects sensitive loads against grid voltage sags without a phase jump. To guarantee the operating stability and good voltage compensation performance of VQC in different sag issues (different duration time of voltage sag or load variation), the operation modes and logical judgment principle are elaborated in detail. Based on the switching scheme, the logic units are defined as:

$$A = \begin{cases} 0 \ \max(U_{VQC}) \le U_{\max} \\ 1 \ \max(U_{VQC}) > U_{\max} \end{cases} \quad \mathrm{B} = \begin{cases} 0 \ ksag \le 1 - \cos(\theta_L) \\ 1 \ ksag > 1 - \cos(\theta_L) \end{cases}$$

Table 2.1 Operation modes of the proposed method

Mode	A	B	C	Details
1	0	0	0	Nonactive power compensation
2	0	1	0	Minimal active power compensation
3	1	0/1	0	Limited voltage compensation
4	0/1	0/1	1	Self-updated operation upon modes 1/2/3

$$C = \begin{cases} 0 \; U_{dc} > U_{dc\,min} \\ 1 \; U_{dc} \le U_{dc\,min} \end{cases} \quad K = \begin{cases} 0 \; ksag \le 0.05 \\ 1 \; ksag > 0.05 \end{cases}. \tag{2.24}$$

The parameters definitions of logical judgment are as follows: K is used to estimate whether a sag issue occurs, B determines that the sag can be corrected by the nonactive power or not, A is used to estimate whether the overmodulation operation of the VQC may happen, C represents the DC link voltage which has already below its minimum value.

Corresponding, the operation modes of the VQC based on the logical judgment mentioned above are as follows.

(1) Mode 1: Nonactive power compensation. As long as the operation boundary of VQC is not exceeded, meanwhile B, C is equal to 0, the nonactive power compensation is activated.
(2) Mode 2: Minimal active power compensation. As long as operation boundaries of VQC is not be exceeded, meanwhile B is equal to 1 and C is equal to 0, the minimal active power compensation is activated.
(3) Mode 3: Limited voltage compensation. Once A is equal to 1, the limited voltage compensation defined as mode 3 is implemented.
(4) Mode 4: Self-updated operation. With the duration of the sag issue, the energy stored in the DC link is reduced due to certain nonzero active power injected by VQC, and consequently, the DC link voltage reduces. To guarantee device stability, updated procedure defined as mode 4 is activated if C is equal to 1 (Table 2.1).

2.4.2 *Overall System Controller Scheme*

Figure 2.9 depicts the detailed block diagram of the proposed comprehensive scheme in the case of phase A. To realize the voltage compensation and elastic recovery, the initial step involves obtaining the reference load voltage, which is divided into two submodules: (1) The logic unit K is employed to confirm that the VQC works in the voltage compensation or recovery stage, the phase calculation block computes the phase angles θ_{trans1} and θ_{trans2}, and the phase of the system voltage θ_0 is considered as the reference and is obtained by PLL. Finally, the VQC injection angle is obtained through the combination of θ_{trans} and θ_0. (2) The reference value of the load voltage is calculated. An imaginary phase variable U_{SaIm}, which lags 90° behind the original phase variable U_{Sa}, is hypothesized, based on the work as presented in reference. The single-phase *d-q* vector control is applied in the controller. Thus, we have

$$\begin{pmatrix} U_d \\ U_q \end{pmatrix} = \begin{pmatrix} \cos(\theta_{\mathrm{VQC}}) & \sin(\theta_{\mathrm{VQC}}) \\ -\sin(\theta_{\mathrm{VQC}}) & \cos(\theta_{\mathrm{VQC}}) \end{pmatrix} \begin{pmatrix} U_{\mathrm{Sa}} \\ U_{\mathrm{SaIm}} \end{pmatrix}, \tag{2.25}$$

The equations in *d-q* coordinates of U_{La}^* are derived as

$$U_{\mathrm{La\_d}}^* = U_{\mathrm{Sa\_}d} = U_d, \tag{2.26}$$

$$U_{\mathrm{La\_}q}^* = \sqrt{\left(U_{\mathrm{La}}^*\right)^2 - \left(U_{\mathrm{Sa\_d}}\right)^2} = \sqrt{\left(U_{\mathrm{La}}^*\right)^2 - \left(U_{\mathrm{q}}\right)^2}. \tag{2.27}$$

Finally, using (2.26) and (2.27), the reference value of U_{La}^* can be obtained through the inverse single-phase d-q transformation [14]. Then, the VQC reference voltage $U_{\mathrm{VQC}}{}^*{}_{\mathrm{a}}$ is obtained and compared with the actual voltage in the stationary reference frame. A PI controller is used for the accurate track of $U_{\mathrm{VQC}}{}^*{}_{\mathrm{a}}$. After unipolar SPWM modulation technology is applied, the driving signal of the IGBT is obtained. This chapter is mainly focused on the mode transition and updated procedure while the pre-sag compensation completed already. Some methods complete detection within 1 ms after the perturbation appearance [37–41], and have good dynamic response even in harmonic environments [39–43]. For the sake of simplicity, the fault detection for the activation of pre-sag compensation is not implemented.

2.5 Simulation Results

In this section, a MATLAB/SIMULINK model with the proposed control scheme, as shown in Fig. 2.1, is built, and its performance is validated in various sag cases. The parameters used in the simulation work are presented in Table 2.2. The performance of the system model is determined using solver 23tb in the variable step with a sample time of 20 μs. The switching frequency used for the IGBT-based VSC is 10 kHz. The value of Δt_1 and Δt_2 are defined as 30 ms. Actually, Δt_1 can be reduced appropriately to minimize the energy consumption on the DC link side, in the case of satisfying the tolerance of phase jump change rate of power equipment. And a detailed elaboration of Δt_1 is given in Appendix.

Table 2.2 Simulation parameters of proposed control scheme

Parameters	Variable	Value
Source parameters	U_S, Z_S, f_0	220 V, 0.05 Ω, 50 Hz
Load parameters	S_{load}, Load PF	1.56 kVA, 0.8
DC link capacitor	C_{dc}	3000 μF
DC link voltage	U_{dc}	360 V
Operation limits	U_{dcmin}	250 V
LC filter inductor	L_f	3 mH
LC filter capacitor	C_f	$C_f = 40$ μF
Switching frequency	f	10 kHz
Transition time	$\Delta t_1, \Delta t_2$	30 ms
Series transformer turn ratio		1:1
Primary voltage		380 V
Secondary voltage		380 V
Power rating		50 kVA
Series transformer	U_k	4%
Magnetization resistance		300 p.u
Magnetization inductance		300 p.u

2.5.1 *Basic Function Verification of Voltage Compensation and Flexible Recovery*

In the first scenario, a sag depth of 20% (meeting the limit given in (2.1)) is considered with a phase jump of 10° for ten cycles and initiated at time $t = 0.30$ s. Using (2.9) and (2.13), the values of $1.414U_{VQCf}$ and $1.414U_{ri}$ are obtained as 186 and 246 V, respectively (lower than the limit given in the (2.18)).

Hence, as seen in Fig. 2.10a, from 0.30 to 0.33 s, the VQC may quickly switch to compensation mode 1 through transition 1. The active–reactive power profiles of VQC-injected are shown in Fig. 2.10b. It can be seen that the controller makes the VQC gradually shift to zero active power compensation and then maintains this state unchanged. Moreover, Fig. 2.10c shows a negligible drop in the DC link voltage (the U_{dc} displayed is reduced tenfold). The value of α_r acquired after the voltage sag removal is 46.86°, and the VQC may quickly recover to standby mode through transition 2, which realizes the seamless connection between the compensation stage and the recovery stage. The amplitudes of U_L and I_L are kept constant with a negligible phase jump during the voltage fluctuations, whether in the compensation or recovery stage, as shown in Fig. 2.10a and c.

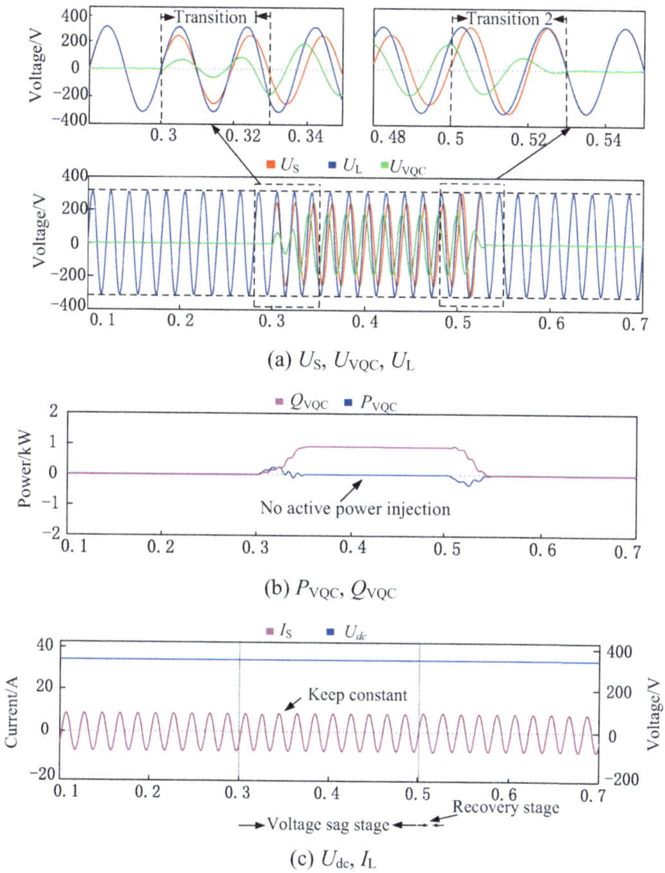

Fig. 2.10 Simulation results of the VQC during voltage fluctuation with k_{sag} is equal to 0.2

2.5.2 Operation Limits and Updated Performance

In the second scenario, a sag depth of 40% (exceeding the limit given in (2.1)) is considered with a phase jump of 10° for twenty cycles and initiated at time $t = 0.30$ s. The values of $1.414U_f$ and $1.414U_{ri}$ are kept at 186 V and 246 V, respectively. Hence, as seen in Fig. 2.11a, from 0.30 s to 0.33 s, the VQC may quickly switch to compensation mode 2 through transition 1. The VQC-injected active–reactive power profiles are shown in Fig. 2.11b. It can be observed that the controller makes the VQC gradually shift to the minimal active power compensation and then maintains this state unchanged. Figure 2.11c shows a constant drop in the DC link voltage. When the DC link voltage drops to the value of Udcmin==250 V at 0.60 s, the updated procedure is initiated using (2.21). The injected voltage then decreases without overmodulation, and the active power profiles of VQC-injected increases, as seen in Fig. 2.11a and b from 0.60 to 0.70 s, respectively. The value of α_r can be acquired after the voltage

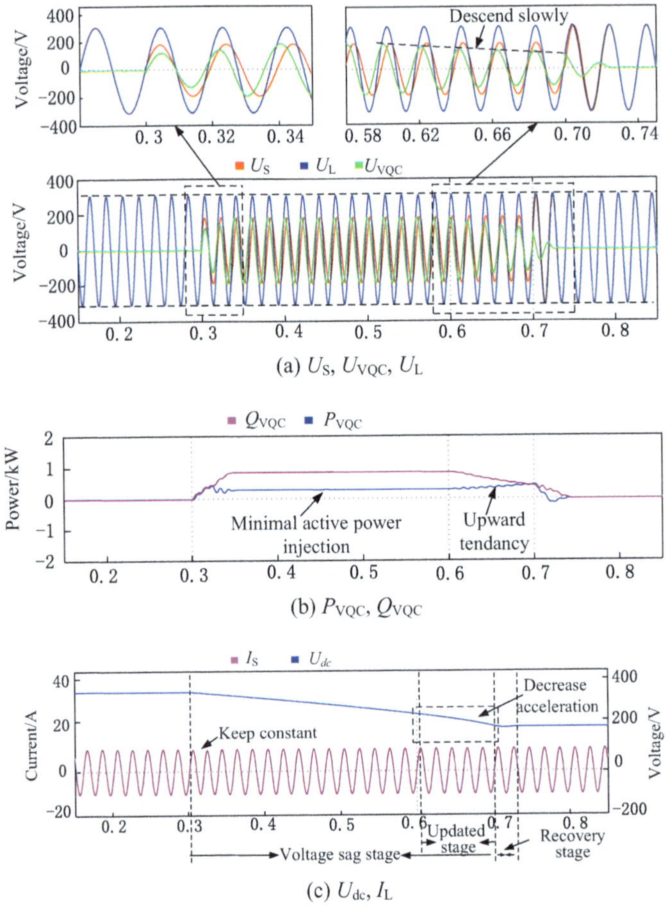

Fig. 2.11 Simulation results of the VQC during the voltage fluctuation with k_{sag} is equal to 0.4

sag removal as 34.72° using (2.19) and (2.23). Then, the recovery stage is initiated at 0.7 s, and an effective link is formed with the final operating point of the compensation stage. The amplitudes of U_L and I_L are still kept constant, as shown in Fig. 2.11a and c.

2.6 Experiment and Analysis

A single-phase experimental verification of the proposed approach was undertaken.

Case1: Minimal active power compensation is activated firstly, then, the self-updated operation mode is activated due to the sag duration.

In the first scenario, a sag depth of 30% (exceeding the limit given in (2.1)) is considered with a phase jump of 10° for twenty cycles and initiated at t_1. The values of $1.414U_f$ and $1.414U_{ri}$ are kept at 189 V and 248 V, respectively. Hence, as seen in Fig. 2.12b, from t_1 to t_2, the VQC may quickly switch to compensation mode 2 through transition 1. It can be observed that the controller makes the VQC gradually shift to the minimal active power compensation and then maintains this state unchanged. Figure 2.12b shows a constant drop in the DC link voltage.

Once the value of load current is doubled at t_3, the energy consumption from DC link is increased, which makes the drop rate increases as well, as seen in Fig. 2.12b. When the DC link voltage drops to the value of $U_{dc\ min}$=250 V at t_4, the updated procedure is initiated using (2.21). The injected voltage decreases without overmodulation, as seen in Fig. 2.12b. Then, the recovery stage is initiated at t_5, and an effective linkage is formed with the final operating point of the compensation stage. The amplitude of U_L is kept constant, as shown in Fig. 2.12a.

Case2: Limited voltage compensation is activated.

In the second scenario, a sag depth of 40% (exceeding the limit given in (2.1)) is considered with a phase jump of 20° for ten cycles and initiated at t_1.

The values of $1.414U_f$ and $1.414U_{ri}$ are obtained as 186 and 295 V (higher than the limit given in (2.18)). Thus, the final operating point during the voltage

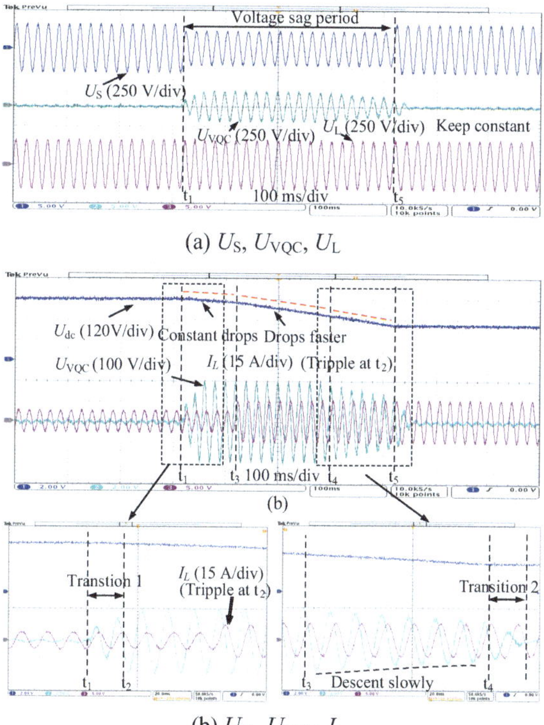

(a) U_S, U_{VQC}, U_L

(b) U_{dc}, U_{VQC}, I_L

Fig. 2.12 Performance of the VQC under a sag depth of 30% with a phase jump of 10° for twenty cycles

Fig. 2.13 Performance of the VQC under a sag depth of 40% with a phase jump of 20° for ten cycles

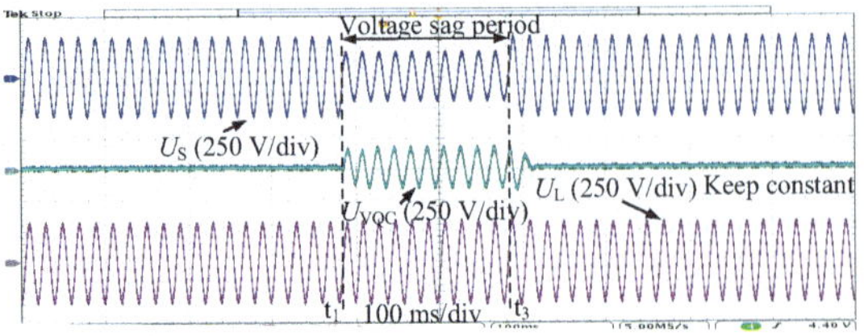

(a) U_S, U_{VQC}, U_L

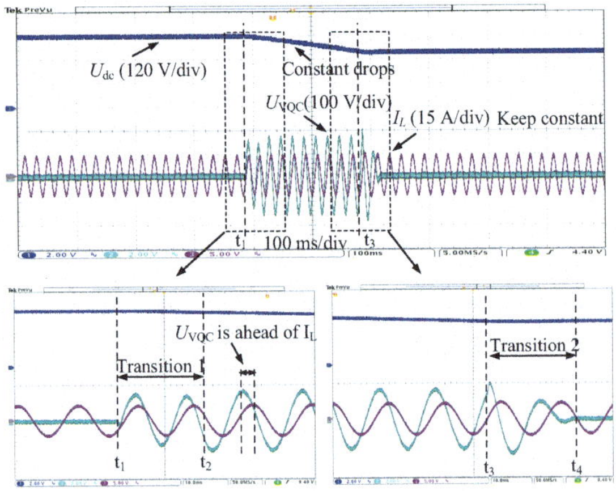

(b) U_{dc}, U_{VQC}, I_L

compensation stage is regulated at $1.414U_{ri} = 250$ V, and α' can be calculated as 47.56° using (2.19). As seen in Fig. 2.13b, from t_1 to t_2, the VQC quickly switches to compensation mode 3 through transition 1. The phase angle of U_{VQC} is ahead of I_L during the stable stage. In this case, the DC link voltage drops in a constant rate and is higher than 250 V until the sag disappears. Thus, the updated procedure is deactivated. The smooth recovery is completed through transition 2 from t_3 to t_4.

Case3: Limited voltage compensation is activated firstly, then, the self-updated operation mode is activated due to the sag duration.

In the third scenario, a sag depth of 40% (exceeding the limit given in (2.1)) is considered with a phase jump of 10° for twenty cycles and initiated at time t_1. The VQC switches to compensation mode 2 through transition 1. Furthermore, the updated procedure of mode 4 is initiated at t_3, where U_{dc} is less than 250 V, the phase angle of U_L ahead of I_L decreases and the DC link voltage drops with a faster fall rate, as seen in Fig. 2.14b. The amplitude of U_L is kept constant with a negligible phase jump in the compensation as well as the recovery stage.

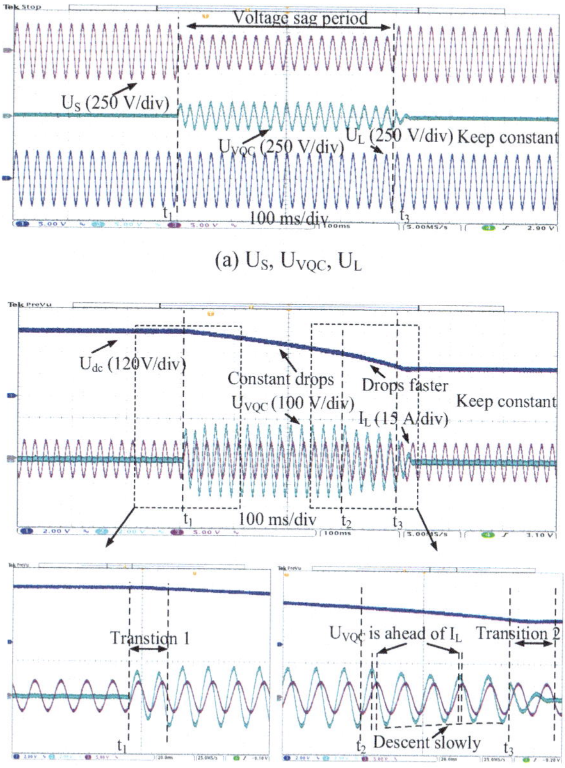

(a) U_S, U_{VQC}, U_L

(b) U_{dc}, U_{VQC}, I_L

Fig. 2.14 Performance of the VQC under a sag depth of 40% with a phase jump of 10° for twenty cycles

Appendix

Relevant parameters can be calculated as follow. Controlling the changing rate of phase angle to satisfy the load tolerance is the first priority, and Δt_1 can be chosen as small as possible on this basis.

In this chapter, once the sag is detected, the first step is the implementation of pre-sag compensation strategy to restore the load voltage and phase to the pre-sag values. Then, once the phase jump is calculated, the controller initiates the transition from pre-sag compensation to energy-optimized strategy. The change rate of phase angle during the transition is calculated as

$$\sigma' = \frac{d\sigma}{dt} = \frac{\sigma_{max}}{\Delta t_1}, \tag{2.A1}$$

where σ_{max} is the maximum phase jump needed to be corrected. Meanwhile, the tolerance of power equipment to the changing rate of phase angle is inequality [44, 45], the value of σ' cannot exceed the tolerable change rate of phase angle σ'_{ref}, which depends on load characteristics. So, we have

$$\Delta t_1 \geq \frac{\sigma_{\max}}{\sigma'_{\text{ref}}}. \tag{2.A2}$$

Further, during the transition from the pre-sag compensation to energy-optimized strategy [14], the maximum active power supplied by the VQC may equal to in-phase compensation given as

$$P_{\text{VQC}\max} = k_{\text{sag}} S_L \cos(\theta_L). \tag{2.A3}$$

At the final operating point of transition stage, the minimum active power supplied by the VQC is given as

$$P_{\text{VQCf}} = (k_{\text{sag}} - (1 - \cos(\theta_L)))S_L \cos(\theta_L). \tag{2.A4}$$

Considering the active power supplied by the VQC is constantly changing during the transition, the energy consumption from DC link is obtained by taking the average power over this period, we have

$$W_1 = \int_0^{\Delta t1} P_{\text{VQC}} t dt \approx \frac{\left(P_{\text{VQC}\max} + P_{\text{VQCf}}\right)}{2} \Delta t_1, \tag{2.A5}$$

where Δt_1 is the time-consuming of the transition from the pre-sag compensation to energy-optimized strategy. Defining the expected drop depth in DC link voltage lower than ΔU_{dcref} during this period, we have

$$W_1 \leq \frac{1}{2} C_{\text{dc}} \left(U_{\text{dc}}^2 - (U_{\text{dc}} - \Delta U_{\text{dcref}})^2\right). \tag{2.A6}$$

Accordingly, substituting (B5) into (B6), Δt_1 can be obtained as

$$\Delta t_1 \leq \frac{C_{\text{dc}}\left(U_{\text{dc}}^2 - (U_{\text{dc}} - \Delta U_{\text{dcref}})^2\right)}{\left(P_{\text{VQCmax}} + P_{\text{VQCf}}\right)}. \tag{2.A7}$$

Overall, the value of Δt_1 is determined by the load characteristics and the capacity of DC link, and can be expressed as

$$\frac{\sigma_{\max}}{\sigma'_{\text{ref}}} \leq \Delta t_1 \leq \frac{C_{\text{dc}}\left(U_{\text{dc}}^2 - (U_{\text{dc}} - \Delta U_{\text{dcref}})^2\right)}{\left(P_{\text{VQCmax}} + P_{\text{VQCf}}\right)}. \tag{A8}$$

References

1. J.G. Nielsen, F. Blaabjerg, A detailed comparison of system topologies for dynamic voltage restorers. IEEE Trans. Indus Appl. **41**(5), 1272–1280 (2005)
2. D. Mahinda Vilathgamuwa, H.M. Wijekoon, S.S. Choi, A novel technique to compensate voltage sags in multiline distribution system—the interline dynamic voltage restorer. IEEE Trans. Ind. Electron. **53**(5), 1603–1611 (2006)
3. S. Biricik, H. Komurcugil, Optimized sliding mode control to maximize existence region for single-phase dynamic voltage restorers. IEEE Trans. Indus. Inform. **12**(4), 1486–1497 (2016)
4. F. Jiang, C. Tu, Q. Guo, Z. Shuai, X. He, J. He, Dual-functional dynamic voltage restorer to limit fault current. IEEE Trans. Ind. Electron. **66**(7), 5300–5309 (2019)
5. T. Kang, S. Choi, A.S. Morsy, P.N. Enjeti, Series voltage regulator for a distribution transformer to compensate voltage sag/swell. IEEE Trans. Indus Electron. **64**(6), 4501–4510 (2017)
6. Somayajula, M.L. Crow, An integrated dynamic voltage restorer ultracapacitor design for improving power quality of the distribution grid. IEEE Trans. Sustain. Energy **6**(2), 616–624 (2015)
7. C. Ho, H. Chung, K. Au, Design and implementation of a fast dynamic control scheme for capacitor-supported dynamic voltage restorers. IEEE Trans. Power Electron. **23**(1), 237–251 (2008)
8. A.M. Massoud, S. Ahmed, P.N. Enjeti, B.W. Williams, Evaluation of a multilevel cascaded-type dynamic voltage restorer employing discontinuous space vector modulation. IEEE Trans. Ind. Electron. **57**(7), 2398–2410 (2010)
9. M.G. Anthony, F. Robinson, W. Yuan, A superconducting magnetic energy storage-emulator/battery supported dynamic voltage restorer. IEEE Trans. Energy Con. **32**(1), 55–62 (2017)
10. Sadigh, K., Smedley, K.M., Review of voltage compensation methods in dynamic voltage restorer (DVR), in *Proceeding IEEE Power and Energy Society General Meeting*, pp. 1–8 (2012)
11. P. Jayaprakash, B. Singh, D.P. Kothari, A. Chandra, K. Al-Haddad, Control of reduced-rating dynamic voltage restorer with a battery energy storage system. IEEE Trans. Ind. Appl. **50**(2), 1295–1303 (2014)
12. H.K. Al-Hadidi, A.M. Gole, D.A. Jacobson, Minimum power operation of cascade inverter-based dynamic voltage restorer. IEEE Trans. Power Deliv. **23**(2), 889–895 (2008)
13. Z. Li, W. Li, T. Pan, An optimized compensation strategy of DVR for micro-grid voltage sag. Protect. Control Mod. Power Syst. **1**(1), 78–85 (2016)
14. A.M. Rauf, V. Khadkikar, An enhanced voltage sag compensation scheme for dynamic voltage restorer. IEEE Trans. Ind. Electron. **62**(5), 2683–2691 (2015)
15. A. Elserougi, A.M. Massoud, A.S. Abdel-Khalik, S. Ahmed, A.A. Hossam-Eldin, An interline dynamic voltage restoring and displacement factor controlling device (IVDFC). IEEE Trans. Power Electron. **29**(6), 2737–2749 (2014)
16. M. Shahabadini, H. Iman-Eini, Improving the performance of a cascaded H-bridge-based interline dynamic voltage restorer. IEEE Trans. Power Del. **31**(3), 1160–1165 (2016)
17. S. Jothibasu, M.K. Mishra, An improved direct AC–AC converter for voltage sag mitigation. IEEE Trans. Ind. Electron. **62**(1), 21–29 (2015)
18. C. Meyer, R.W. Doncker, Y.W. Li, F. Blaabjerg, Optimized control strategy for a medium-voltage DVR—theoretical investigations and experimental results. IEEE Trans. Power Electron. **23**(6), 2746–2754 (2008)
19. S.S.H. Bukhari, S. Atiq, B. Kwon, A sag compensator that eliminates the possibility of inrush current while powering transformer-coupled loads. IEEE J. Emerg. Select. Topics Power Electron. **5**(2), 891–899 (2017)
20. S. Subramanian, M.K. Mishra, Interphase AC–AC topology for voltage sag supporter. IEEE Trans. Power Electron. **25**(2), 514–518 (2010)
21. P. Li, L. Xie, J. Han, S. Pang, P. Li, New decentralized control scheme for a dynamic voltage restorer based on the elliptical trajectory compensation. IEEE Trans. Ind. Electron. **64**(8), 6484–6495 (2017)

22. H. Hafezi, R. Faranda, Dynamic voltage conditioner, a new concept for smart low-voltage distribution system. IEEE Trans. Power Electron. https://doi.org/10.1109/TPEL.2017.2772845
23. X. Liu Yingying, X.X. Yonghai, Z. Yongqiang, G. Chunlin, A stages compensation and control strategy for series power-quality regulator. IEEE Trans. Power Del. **25**(4), 2807–2815 (2010)
24. Jian Ye, H.B. Gooi, F. Wu, Optimal design and control implementation of UPQC based on variable phase angle control method. IEEE Trans. Ind. Inf. https://doi.org/10.1109/TII.2018.2834628
25. M. Pradhan, M.K. Mishra, Dual P-Q theory based energy optimized dynamic voltage restorer for power quality improvement in distribution system. IEEE Trans. Ind. Electron., Citation Inf. https://doi.org/10.1109/TIE.2018.2850009
26. J. Wang, Y. Xing, T. Yang, A novel dual-DC-port dynamic voltage restorer with reduced-rating integrated DC-DC converter for wide-range voltage sag compensation. IEEE Trans. Power Electron. https://doi.org/10.1109/TPEL.2018.2882534
27. J.G. Nielsen, M. Newman, H. Nielsen, F. Blaabjerg, Control and testing of a dynamic voltage restorer (DVR) at medium voltage level. IEEE Trans. Power Electron. **19**(3), 806–813 (2004)
28. F.B. Ajaei, S. Afsharnia, A. Kahrobaeian, S. Farhangi, A fast and effective control scheme for the dynamic voltage restorer. IEEE Trans. Power Del. **26**(4), 2398–2406 (2010)
29. G. Chen, L. Zhang, R. Wang, L. Zhang, X. Cai, A novel SPLL and voltage sag detection based on LES filters and improved instantaneous symmetrical components method. IEEE Trans. Power Del. **30**(3), 1177–1189 (2015)
30. L. Haichun, X. Lizhi, X. Shaojun, A new detection method of voltage sag based on period phase concept, in *Proceeding IEEE Conference on Industrial Electronics and Applications*, China, pp. 3333–3337 (2009)
31. L. Xiong, F. Zhuo, X. Liu, M. Zhu, A fast orthogonal signal-generation algorithm characterized by noise immunity and high accuracy for single-phase grid. IEEE Trans. Power Electron **31**(3), 1847–1851 (2016)
32. F. Xiao, L. Dong, L. Li, X. Liao, A novel open-loop frequency estimation method for single-phase grid synchronization under distorted conditions. IEEE J. Emerg. Select. Topics Power Electron **5**(3), 1487–1493 (2017)
33. J. Roldan-Pérez, A. García-Cerrada, M. Ochoa-Gimenez, J.L. Zamora-Macho, Delayed-signal-cancellation-based sag detector for a dynamic voltage restorer in distorted grids. IEEE Trans. Sustain. Energy. https://doi.org/10.1109/TSTE.2018.2877505
34. Karthikeyan, D.G. Abhilash Krishna, S. Kumar, B. Venkatesa Perumal, S. Mishra, Dual role CDSC based dual vector control for effective operation of DVR with harmonic mitigation. IEEE Trans. Ind. Electron. https://doi.org/10.1109/TIE.2018.2829678
35. T.A. Naidu, S.R. Arya, R. Maurya, Dynamic voltage restorer with quasi newton filter based control algorithm and optimized values of PI regulator gains. IEEE J. Emerg. Select. Topics Power Electron. https://doi.org/10.1109/JESTPE.2018.2890415
36. S. Biricik, H. Komurcugil, N.D. Tuyen, M. Basu, Protection of sensitive loads using sliding mode controlled three-phase DVR with adaptive notch filter. IEEE Trans. Ind. Electron. https://doi.org/10.1109/TIE.2018.2868303
37. P. Thakur, A.K. Singh, R.C. Bansal, Novel way for classification and type detection of voltage sag. IET Gen. Trans. Distrib. **7**(4), 398–404 (2013)
38. K. Sadigh, K.M. Smedley, Fast voltage sag detection method for single-/three-phase application, in *Proceeding IEEE APEC*, pp. 881–888 (2013)
39. J.J. Bae, J. Lee, B. Han, Novel sag detection method for line interactive dynamic voltage restorer. IEEE Trans. Power Del. **25**(2), 1210–1211 (2010)
40. Y. Kumsuwan, Y. Sillapawicharn, A fast synchronously rotating reference frame-based voltage sag detection under practical grid voltages for voltage sag compensation systems. Proc. Int. Conf. PEMD, pp. 1–5 (2012)
41. F. Xiao, L. Dong, L. Li, X. Liao, Fast voltage detection method for grid-tied renewable energy generation systems under distorted grid voltage conditions. IET Power Electronics **10**(12), 1487–1493 (2017)

42. Y. Han, M. Luo, X. Zhao, J.M. Guerrero, X. Lin, Comparative performance evaluation of orthogonal signal generators based single-phase PLL algorithms—a survey. IEEE Trans. Power Electron. **31**(5), 3932–3945 (2016)
43. A. Roshan, R. Burgos, A.C. Baisden, F. Wang, D. Boroyevich, A D-Q frame controller for a full-bridge single phase inverter used in small distributed power generation systems, in *Proceeding IEEE APEC, USA*, pp. 641–647 (2007)
44. M. Mansour, S. Islam, M.A.S. Masoum, Impacts of symmetrical and asymmetrical voltage sags on DFIG-based wind turbines considering phase-angle jump, voltage recovery, and sag parameters. IEEE Trans. Power Electron. **26**(5), 1587–1598 (2011)
45. L. Bing, Effect of voltage sag phase factors on the performance of asynchronous motor, in *Proceeding IEEE 2017 29th CCDC*, pp. 3218–3233 (2017)

Chapter 3
The Output Harmonic Minimization Strategy for Voltage Quality Controllers

This chapter outlines an optimized control methodology designed to enhance the output voltage level count of cascaded H-bridge voltage quality controllers (CHB-VQC). To validate the viability of this approach, both conventional in-phase compensation and optimal control strategies are employed to assess the quality of the output voltage. The subsequent simulation and experimental findings underscore the proficiency and effectiveness of the proposed control scheme in achieving its objectives.

3.1 Introduction

A variety of multilevel inverter designs have been introduced to assure that the CHB-VQC can produce the necessary compensatory voltage. However, a significant concern arises from the intimate connection between the number of output voltage levels produced by the CHB-VQC (Fig. 3.1) and the degree of voltage sag experienced [1]. Specifically, minor sags result in a reduced modulation index, leading to a lower number of steps in the output voltage. Conversely, more severe sags induce a higher modulation index and a greater number of voltage steps. This variability means that the conventional CHB-VQC is unable to consistently provide a compensatory voltage at its maximum level number across various sag depths.

To tackle this limitation, a scheme utilizing a DC-DC converter was presented in Ref. [2], which modifies the modulation index by adjusting the VQC's DC link voltage. This allows the inverter's output voltage to remain consistently at its peak level, regardless of the sag's intensity. In Ref. [3], a mathematical formulation is proposed to calculate the distortion at the point of common coupling, enabling the selection of the appropriate level number for differing circumstances. Reference [4] introduces an innovative adjustable DC link voltage architecture that incorporates a solar energy conversion system. While this adjustment is intriguing, it necessitates

F. Jiang et al., *Advanced Voltage Quality Controller for New Power Distribution Systems*, https://doi.org/10.1007/978-981-97-9861-2_3

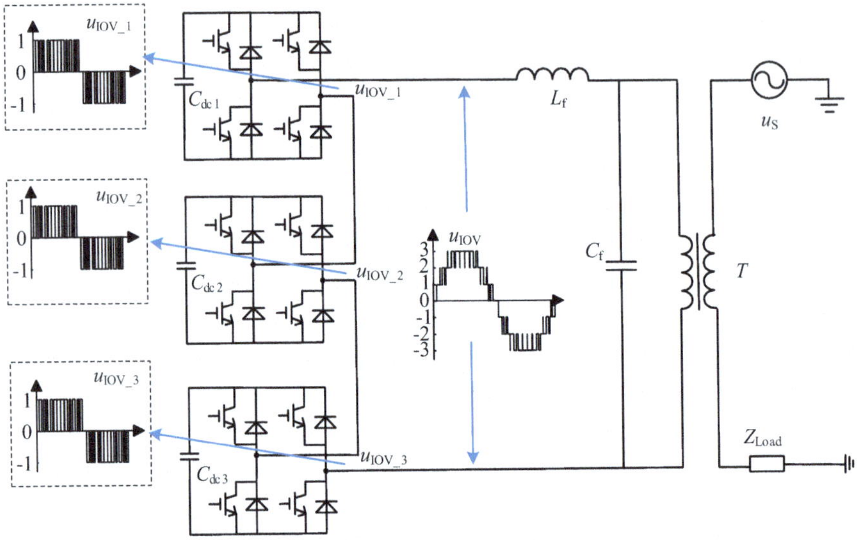

Fig. 3.1 Schematic of a CHB-VQC system

a DC link voltage that is largely constant [5]. Furthermore, frequent alterations to the DC link voltage could potentially compromise system security and stability [6]. Another approach discussed in Ref. [7] involves making the transformer's turns ratio adjustable, thereby providing a novel means of modifying the input voltage's amplitude.

In this endeavor, an optimized control strategy is introduced to maximize the number of output voltage levels from the CHB-VQC, ensuring the delivery of a high-quality voltage with minimized THD. The core of our proposed methodology revolves around dynamically adjusting the reference voltage for compensation in accordance with the depth of the voltage sag. This approach circumvents the need for frequent modifications to the DC link voltage, thereby alleviating potential concerns regarding system stability and security. Consequently, the CHB-VQC demonstrates notable effectiveness, particularly when dealing with voltage sags of lesser magnitude. This optimized control scheme not only enhances the adaptability of the CHB-VQC across a spectrum of sag conditions but also ensures that the system operates efficiently and reliably, even in scenarios characterized by shallow voltage sags where traditional approaches might struggle to maintain the desired level of voltage quality.

3.2 Topology and Compensation Scheme

3.2.1 Topology

Figure 3.2 depicts a single-phase configuration of a CHB-VQC. This setup comprises three principal components: the input module, the DC link module, and the output module. Within the input module, three parallel inverters are interconnected with the grid via a multi-winding transformer. Each of these inverters is coupled with a filter (L_z) designed to attenuate high-frequency ripples.

The output module features a seven-level cascaded inverter that interfaces with the grid through a series transformer (N_a) and an LC filter. This module and the input module are linked by a C_{dc}, which plays a pivotal role in stabilizing the DC link voltage. The DC link module serves as the energy reservoir for compensation purposes, while the output module is responsible for injecting the compensatory voltage into the system, post-filtering through the LC filter.

The primary function of the CHB-VQC is to mitigate the effects of voltage sags, ensuring a stable and reliable power supply. The injection voltage is expressed

$$u_{\text{vqc}} = u_{\text{Load}} - u_{\text{s}}, \tag{3.1}$$

where u_{vqc} is the compensation voltage, u_{Load} is the rated load voltage (the reference load voltage), and u_S is the source voltage during sags. In the present study, phase-shifted carrier pulse width modulation (PSC–PWM) is used in the CHB-VQC [8, 9], and the output fundamental voltage value is expressed as

$$u_{\text{vqc}} = Nmu_{\text{dc}} \sin(\omega_s t + \theta_s), \tag{3.2}$$

where N is the CHB module number, m is the modulation index, u_{dc} is the DC link voltage for each module, ω_s is the angular frequency and θ_s is the starting phase angle. In (3.2), the amplitude of the inverter fundamental output voltage is related to N, m, and u_{dc}. The modulation index is calculated as

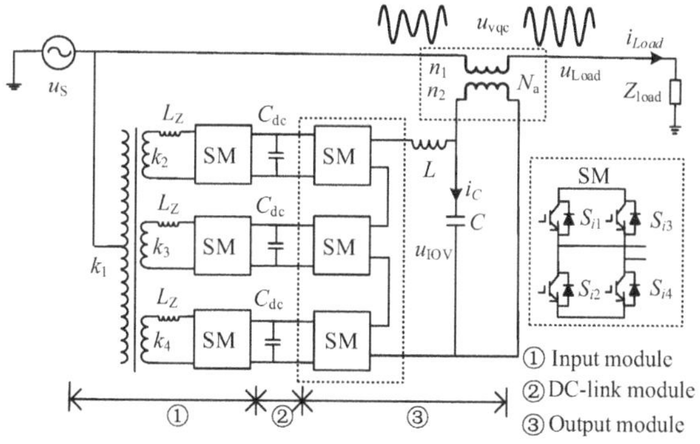

Fig. 3.2 Topology of the CHB-VQC system

$$m = \frac{u_{\text{vqc}}}{N u_{\text{dc}}}. \tag{3.3}$$

In (3.3), when N is selected, two methods are observed to increase the modulation index m. The first method decreases u_{dc}, and the second increases u_{vqc}. However, the former may easily worsen the CHB-VQC stability given the frequent DC link voltage fluctuations [10]. Thus, the latter is selected for the present study.

3.2.2 Compensation Scheme

The voltage compensation scheme of the CHB-VQC mainly includes three methods, namely the in-phase, phase advance, and minimum energy compensation schemes [11, 12]. In Fig. 3.3, 'O' is the center of the amplitude trajectory circle of the load voltage (dotted circle), u_{pre} is the load voltage before the sag, u_S is the system voltage after the sag, i_L is the load current after the sag, $\Delta\theta$ is the phase jump angle, and u_{vqc} is the amplitude of the output compensation voltage. For the first two schemes, the amplitudes of the compensation voltage u_{vqc} are constant when u_{s} is detected, as depicted in Fig. 3.3a and b.

However, the minimum energy compensation scheme retains a constant load voltage by minimizing the active power exchange between the DC link module and the external power system. In Fig. 3.3c, the solid circle is the compensation limited range of the CHB-VQC, its center (Point 'C') is the end of u_S, φ_L is the phase angle of the load, u_{ref} is the compensation reference voltage, β is the phase angle of the compensation reference voltage, and φ_{vqc} is the phase angle of output voltage.

β increases (decreases) when u_{ref} has a clockwise (anticlockwise) rotation. Moreover, the fundamental amplitude of the inverter output voltage can be changed flexibly within a certain range by adjusting β [13]. In Fig. 3.3c, the dotted circle and its intersection region (Arc AB) are the adjustment ranges of u_{ref}, in which β can vary. In ΔDOC, the cosine theorem can be derived as

$$\angle \text{DOC} = \arccos \frac{|u_{\text{ref}}|^2 + |u_s|^2 - |u_{\text{vqc}}|^2}{2 \cdot |u_{\text{ref}}| \cdot |u_s|}. \tag{3.4}$$

In Fig. 3.3c, $\angle\text{DOC} = \beta - \Delta\theta$ can be derived. When u_{ref} is rotated to any position on Arc AB, it can be obtained as follow

$$u_{\text{vqc}}^2 = u_{\text{ref}}^2 + u_{\text{s}}^2 - 2u_{\text{ref}}u_{\text{s}}\cos(\beta - \Delta\theta), \tag{3.5}$$

$$u_{\text{vqc}} = \sqrt{u_{\text{ref}}^2 + u_{\text{s}}^2 - 2u_{\text{ref}}u_{\text{s}}\cos(\beta - \Delta\theta)}. \tag{3.6}$$

Equation (3.6) indicates that the compensation amplitude u_{vqc} of the CHB-VQC output voltage can be changed by rotating the phase angle β of the load reference

Fig. 3.3 Phase diagram of the different compensation schemes during sags

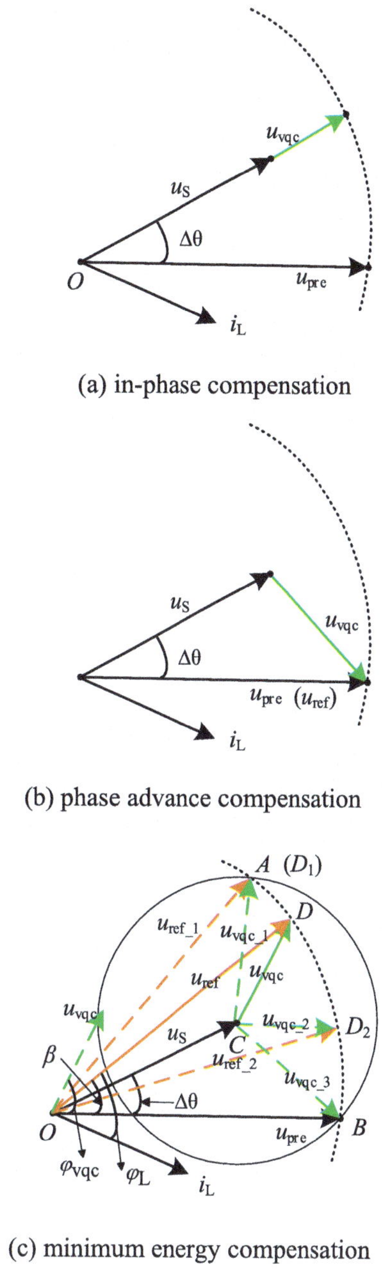

voltage u_{ref}. The amplitude range of u_{vqc} is controlled by β. Therefore, the modulation ratio can be increased by adjusting β without changing the DC link voltage.

3.3 Maximum Level Number Scheme

When the modulation ratio is flexibly controlled, the maximum level number of the compensation voltage is easily achieved [14]. In this section, the maximum level number scheme of the output voltage is introduced in detail.

3.3.1 Relationship Between Output Voltage Level and Modulation Index

In this study, the unipolarity PSC–PWM is selected because of its high DC link voltage utilization. In Fig. 3.4, the modulator signal is $f(t) = m\sin(\omega_s t)$, and the peak values of the triangle carriers are 1. T_{ri1} controls S_{ij}, and T_{ri2} controls S_{ik}, where $i = 1, 2, \ldots, n$; $j = 1, 2$; and $k = 3, 4$. The phase difference between T_{ri1} and T_{ri2} is π, and $T_{r(i+1)1}$ leads to T_{ri1} with π/n. In Figs. 3.2 and 4.4, S_{i1} turns on when $f(t) > T_{ri1}$. By contrast, S_{i2} turns off when $f(t) < T_{ri1}$. And S_{i4} turns on when $f(t) > T_{ri2}$, whereas S_{i3} turns off when $f(t) < T_{ri2}$.

In Fig. 3.4, θ_1' is the intersection point of $T_{\mathrm{r}11}$ and $f(t)$, while θ_i is the intersection point of T_{ri2} and $f(t)$. PWM theory posits that θ_1' and θ_i are the voltage phases of the starting and ending times, respectively. When each CHB module has a high output voltage level number, the phase difference is

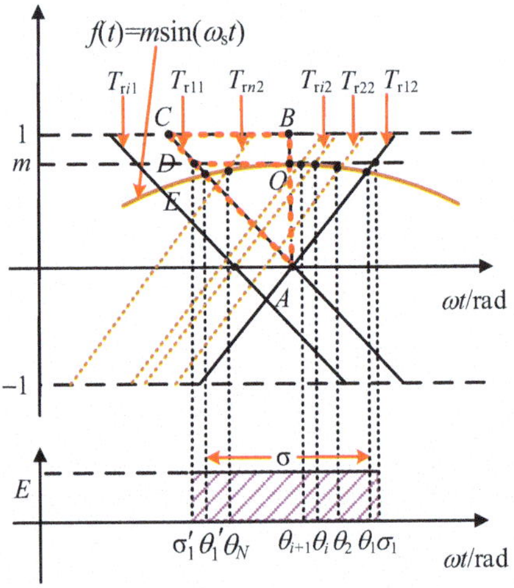

Fig. 3.4 Diagram of the PSC–PWM

$$\sigma = \theta_1 - \theta_1'. \tag{3.7}$$

After the PWM natural sampling [11] is selected, the (3.7) can be rewritten as

$$\sigma \approx \sigma_1 - \sigma_1'. \tag{3.8}$$

From the similar triangle principle in ΔABC, it can be obtained as

$$\frac{\text{AO}}{\text{AB}} = \frac{m}{1} = \frac{\text{OD}}{\text{BC}} = \frac{\sigma/2}{T_{\text{cs}}/4} = \frac{2\sigma}{T_{\text{cs}}}, \tag{3.9}$$

where T_{CS} is the carrier period. PSC-PWM is usually applied to the drive pulse for the CHB inverter, and the shifted phase of the triangular carrier for cell x relative to that for cell 1 is expressed as [15–18]

$$\theta_x - \theta_1 = \frac{(x-1)}{N}\pi, \tag{3.10}$$

where $x = 1, \ldots, N$. Using two-cell CHB as an example, the shifted phases of the triangular carriers for cells 1 and 2 are 0° and 90°, respectively. Moreover, when $x = N$, (3.11) is derived from (3.10) as

$$\theta_N - \theta_1 = \frac{(N-1)}{N}\pi. \tag{3.11}$$

To obtain the $(2N + 1)$ level, $\theta_n \in \left(\sigma_1', \sigma_1\right)$.. Equation (3.12) is derived from (3.7), (3.8) and (3.11) [19]

$$\frac{N-1}{N}\pi < \sigma = \frac{m}{2}T_{\text{cs}}. \tag{3.12}$$

Considering that $T_{\text{CS}} = 2\pi$ in (3.12), we have

$$m > \frac{N-1}{N}. \tag{3.13}$$

3.3.2 Principle of the Maximum Level Output

In this study, u_{vqc} is defined as

$$u_{\text{vqc}} = \text{Nmu}_{\text{dc}}. \tag{3.14}$$

To obtain the maximum level, (3.3) is substituted into (3.12), and $m < 1$ and $T_{CS} = 2\pi$ are considered. Thus,

$$\theta = \arccos \frac{(R_s + R_r/s) + R_{\text{inv}}}{\sqrt{[(R_s + R_r/s) + R_{\text{inv}}]^2 + (jX_s + jX_r + jX_{\text{inv}} + jX_V)^2}}. \tag{3.15}$$

Considering (3.5) and (3.15), if the inverter outputs the maximum level, then the compensation voltage must satisfy

$$(N - 1)u_{\text{dc}} < \sqrt{u_{\text{ref}}^2 + u_{\text{s}}^2 - 2u_{\text{ref}}u_{\text{s}} \cos(\beta - \Delta\theta)} < \text{N}u_{\text{dc}}. \tag{3.16}$$

Furthermore, (3.16) is simplified as

$$\beta_1 < \beta < \beta_2. \tag{3.17}$$

where the variable

$$\beta_1 = \Delta\theta + \arccos \frac{u_{\text{ref}}^2 + u_s^2 - (N - 1)^2 u_{\text{dc}}^2}{2u_{\text{ref}}u_s}, \tag{3.18}$$

$$\beta_2 = \Delta\theta + \arccos \frac{u_{\text{ref}}^2 + u_s^2 - N^2 u_{\text{dc}}^2}{2u_{\text{ref}}u_s}. \tag{3.19}$$

From (3.16)–(3.19), when the voltage sags (u_S) is detected, $\Delta\theta$ should be calculated. Considering that u_{dc} and N are chosen, β is only related to u_{ref}. Therefore, the amplitude of the inverter output voltage can be controlled to achieve the maximum level by adjusting β.

3.4 Improvement Maximum Level Number Scheme

3.4.1 Minimum Energy Angle Scheme

In Fig. 3.3c, the single-phase CHB-VQC outputs the active power as follows:

$$p_{\text{vqc}} = u_{\text{ref}} i_L \cos \varphi_L - u_s i_L \cos \varphi_S. \tag{3.20}$$

If $p_{\text{vqc}} > 0$, then the CHB-VQC outputs the active power; if $p_{\text{vqc}} < 0$, then the CHB-VQC absorbs the active power; and if $p_{\text{vqc}} = 0$, then the CHB-VQC has no active power exchange with the power network. In (3.20), if φ_L, u_{ref} and i_L are constant values, then the active exchange between the CHB-VQC and the system depends on the magnitudes of u_S and φ_s.

If $p_{\mathrm{vqc}} = 0$, then the CHB-VQC has no active power exchange with the system. Simultaneously, (3.20) is rewritten as

$$\varphi_S = \arccos\left(\frac{u_{\mathrm{ref}} \cos \varphi_L}{u_S}\right), \tag{3.21}$$

where $\varphi_s = \varphi_L + \Delta\theta - \beta$. Thus, the following equation is achieved:

$$\beta = \varphi_L + \Delta\theta - \arccos\left(\frac{u_{\mathrm{ref}} \cos \varphi_L}{u_S}\right), \tag{3.22}$$

where $|u_{\mathrm{ref}}\cos\varphi_{\mathrm{L}}/u_{\mathrm{s}}|<1$. Thus, using (3.20), the following equation can be obtained as

$$\frac{\partial p_{\mathrm{vqc}}}{\partial \beta} = -u_s i_L \sin(\varphi_L + \Delta\theta - \beta). \tag{3.23}$$

If $\partial p_{\mathrm{vqc}}/\partial \beta = 0$, then p_{vqc} will be the smallest value. At this time,

$$\varphi_L + \Delta\theta - \beta_0 = 0. \tag{3.24}$$

Using (3.22) and (3.23), the amplitude and phase angle of the compensation voltage can be determined as

$$u_{\mathrm{vqc}} = \sqrt{u_s^2 + u_{\mathrm{ref}}^2 - 2u_s u_{\mathrm{ref}} \cos(\beta_0 - \Delta\theta)}, \tag{3.25}$$

$$\varphi_{\mathrm{vqc}} = \beta_0 + \arccos\left(\frac{u_{\mathrm{vqc}}^2 + u_{\mathrm{ref}}^2 - u_s^2}{2u_{\mathrm{vqc}}u_{\mathrm{ref}}}\right). \tag{3.26}$$

3.4.2 *Improvement Minimum Energy Angle Scheme Based on Maximum Level Output*

When the minimum energy angle β_0 does not satisfy (3.17), an improvement minimum energy angle is adopted. In Fig. 3.3c and (3.17), changing the load reference voltage phase is helpful in moving β_0 to β_1, β_2. Then, the CHB-VQC can output the maximum level. Two cases are available for the selecting β_0.

Case 1: When $\beta_1 < \beta_0 < \beta_2$, p_{vqc} is consistently larger than 0. That is, the CHB-VQC will output the active power. Thus, the minimum power output value can be calculated as follow

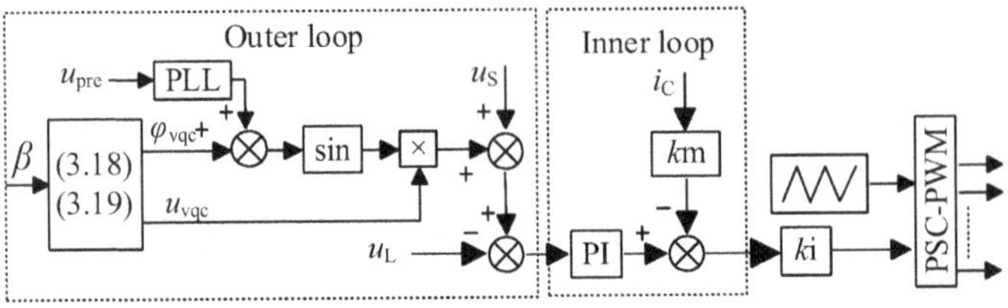

Fig. 3.5 Control block diagram of the CHB-VQC

$$\beta_0' = P^{-1}\left[\min(p(\beta_1), p(\beta_2))\right]. \tag{3.27}$$

Case 2: When $\beta_1 < \beta_0 < \beta_2$, p_{vqc} is constantly smaller than 0. That is, the CHB-VQC will absorb the active power. Thus, the maximum power can be calculated as

$$\beta_0'' = p^{-1}\left[\max(p(\beta_1), p(\beta_2))\right]. \tag{3.28}$$

Utilizing Eqs. (3.27) and (3.28), the amplitude and phase of the compensation voltage are recalculated to optimize the compensation voltage output. In essence, when a voltage sag occurs in the system voltage supply, the power source must maximize its active power output, while the CHB-VQC minimizes its active power output. To ensure maximum output voltage levels, the load reference voltage vector (u_{ref}) is gradually rotated toward the point of minimum value, where the CHB-VQC supplies active power based on u_{pre}. Conversely, if u_{ref} rotates toward the maximum point, the CHB-VQC absorbs active power according to u_{pre}. The control strategy for voltage compensation employs an outer loop of instantaneous load voltage feedback and an inner loop of instantaneous filter capacitor current feedback. Figure 3.5 illustrates the control block diagram for this voltage compensation control strategy.

3.4.3 *Operation Sequence Analysis*

The flowchart of the proposed control scheme is displayed in Fig. 3.6, where d_{sag} is the drop amplitude of the grid voltage. If $d_{sag} \geq 5\%$, grid voltage sag is detected, and the CHB-VQC outputs the compensation voltage. Otherwise, the compensation voltage cannot be generated. Then, φ_S and $\Delta\theta$ are detected, and β_0 is calculated using (3.23) and (3.24). Equation (3.16) is used to determine if β_0 satisfies the maximum output level condition. If the condition is satisfied, then β does not require rotation. If the condition is unsatisfied, then β must be slowly rotated until the requirements of (3.27) and (3.28) are satisfied. Accordingly, the compensation voltage is generated.

Fig. 3.6. Flowchart of the phase adjustment of the load reference voltage

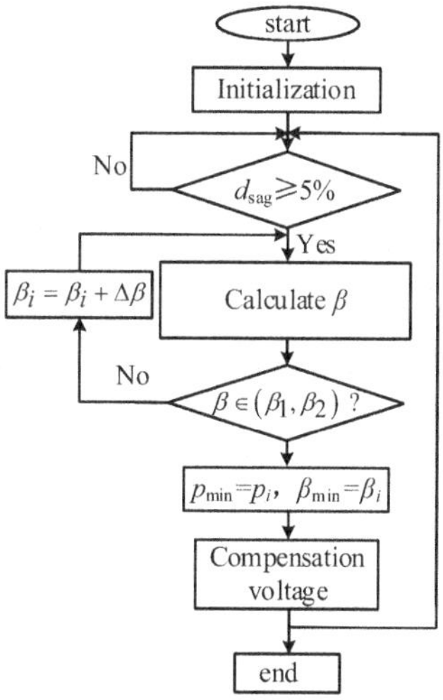

3.5 Simulation Results

In this segment, the simulation outcomes are furnished to substantiate the efficacy and accuracy of the proposed optimal control strategy. The simulations, pertaining to the circuit model depicted in Fig. 3.2, have been executed within the MATLAB environment. A comprehensive enumeration of the simulation parameters utilized is provided in Table 3.1.

Table 3.1 Simulation parameters of the CHB-VQC

Parameters	Variable	Value
System source	u_S	311 V
System impedance	Z_S	0.02 Ω
Switching frequency	–	6.4 kHz
DC link capacitor	C_{dc}	5000 μF
DC module voltage	U_{dc}	40 V
LC filter	L_f, L_C	1.5 mH, 40 μF
Series transformer	n_1:n_2	1:1
Multi-winding transformer	k_1:k_2:k_3:k_4	2:1:1:1
Filter inductor	L_z	0.6 mH
Number of CHBs	–	3
Load 1	Z_{Load1}	23.325 + j20.567 Ω
Load 2	Z_{Load2}	23.325 + j78.5 Ω

3.5.1 Simulation Results with Using Traditional In-Phase Compensation Strategy

Figures delineating the operational proficiency of the CHB-VQC employing an in-phase compensation tactic are presented in Fig. 3.7.

During the initial phase from 0.10 to 0.55 s, the load exhibits an impedance of 23.325 + j20.567 Ω, transitioning to 23.325 + j78.5 Ω in the subsequent phase from 0.55 to 1.00 s. Figure 3.7a showcases that the source voltage endures four distinct sags, with a reduction to 25% of its nominal value from 0.2–0.3 and 0.6–0.7 s, and a shallower dip to 5% of its nominal value from 0.4–0.5 and 0.8–0.9 s. It is evident from Fig. 3.7a that the CHB-VQC successfully introduces a compensatory voltage, denoted as u_{vqc}, to maintain the constancy of the load voltage, u_{Load}, amidst these voltage disturbances.

Furthermore, Fig. 3.7b elucidates that the in-phase compensation strategy facilitates the CHB-VQC's provision of both active and reactive power outputs in response to the varying sags. This figure also reveals that, under the conventional methodology, the CHB-VQC necessitates active power supply from the DC link across different sag events.

Figure 3.7c demonstrates that the compensation voltage level, designated as u_{IOV}, ascends to 7 when the sag depth reaches 25% of the nominal voltage level. Conversely, during a 5% sag, the u_{IOV} level diminishes to 3. This observation underscores the adaptability of the output voltage level in accordance with the fluctuating depths of the voltage sags.

Additionally, upon transitioning from the first to the second stage, the load's alteration does not precipitate any variation in the output voltages of the CHB-VQC during the sag occurrences. Figure 3.7d attests to the seamless transition of the load current, devoid of any surges or disruptions, as it navigates between the distinct stages. This characteristic highlights the system's capability to ensure smooth operation despite dynamic load conditions.

Figure 3.8 illustrates the variation in harmonic content across different voltage sag conditions. As depicted in Fig. 3.8a, the CHB-VQC is capable of dynamically generating distinct compensation voltages during the intervals of 0.2–0.3 s, 0.4–0.5 s, 0.6–0.7 s, and 0.8–0.9 s. However, the figure also reveals that the THD of the compensation voltages is 1.46 and 1.51% during the deeper sag periods of 0.2–0.3 s and 0.6–0.7 s, respectively.

Contrastingly, Fig. 3.8c shows that during the shallower sag periods of 0.4–0.5 s and 0.8–0.9 s, the THDs of the compensation voltages notably increase to 6.65% and 6.70%, respectively. This observation, highlighted in Fig. 3.8a, suggests that shallower voltage sags lead to a higher THD in the compensation voltage, indicating that the performance of the CHB-VQC may be suboptimal when dealing with such conditions.

Therefore, it becomes apparent that the CHB-VQC encounters limitations in effectively managing the quality of the compensation voltage during shallow voltage sags,

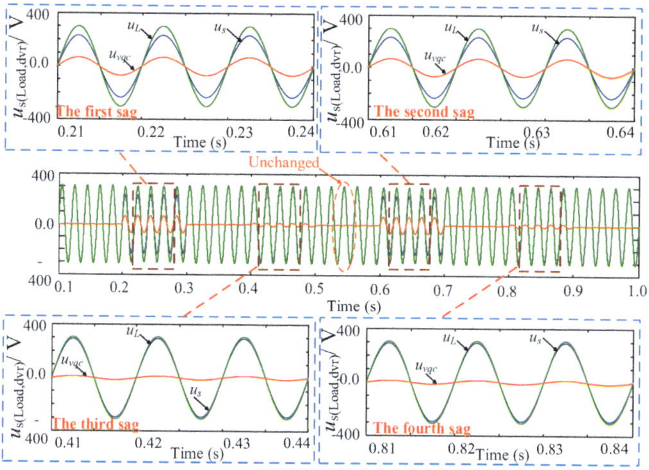

(a) source voltage u_S, compensation voltage u_{vqc} and load voltage u_{Load}

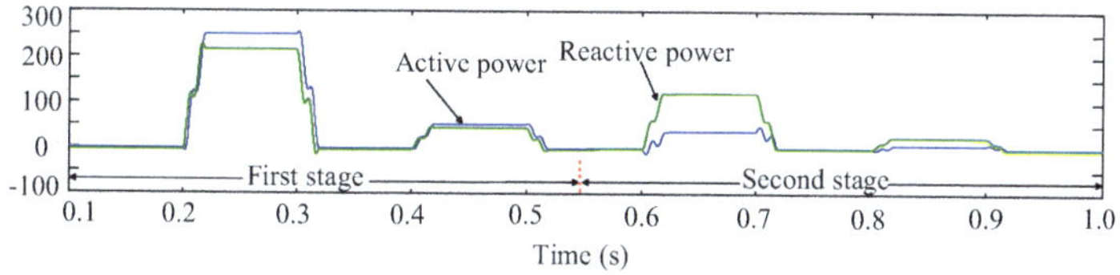

(b) active power p_{vqc} and reactive power q_{vqc}

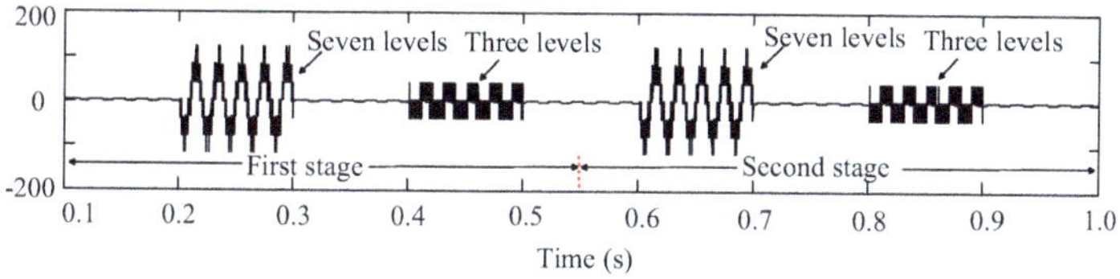

(c) stepped output voltage of the inverter u_{IOV}

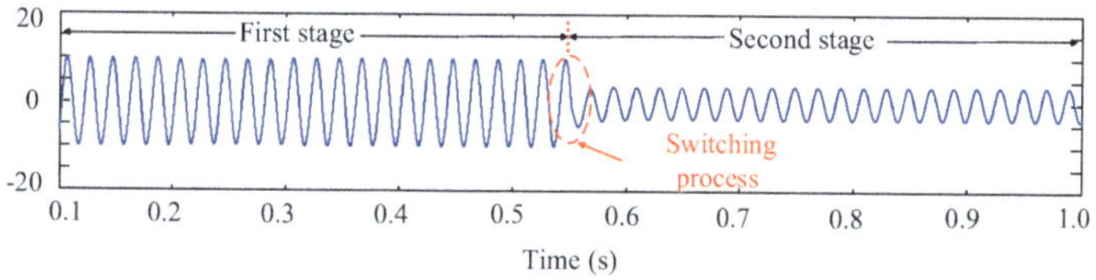

(d) load current i_{Load}

Fig. 3.7 Simulation results of the CHB–VQC under in-phase compensation

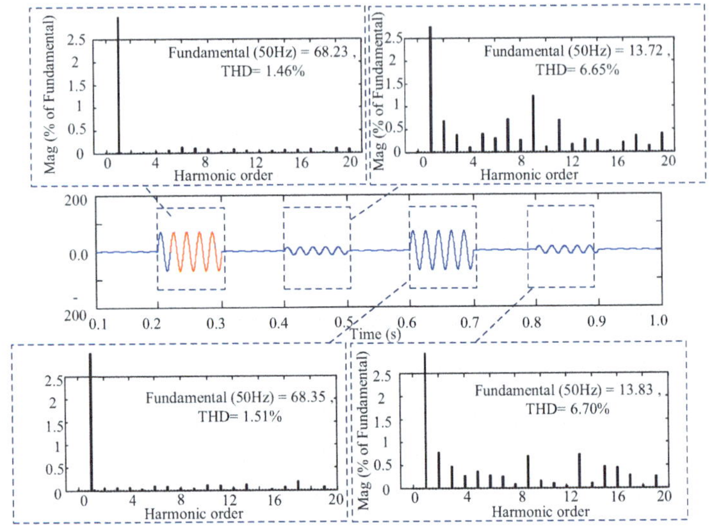

(a) compensation voltage u_{vqc}

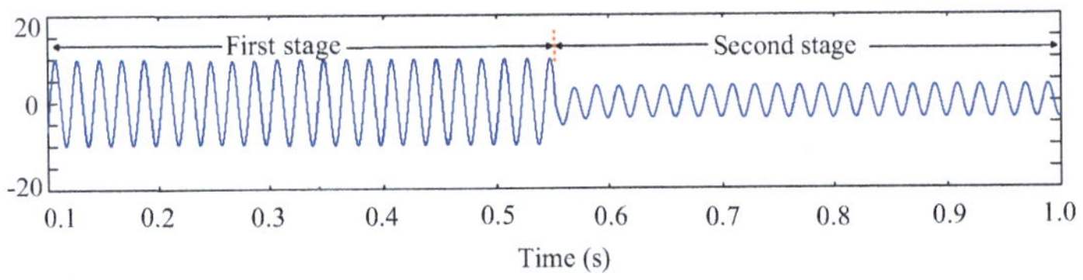

(b) load current i_{Load}

Fig. 3.8 Harmonic content under in-phase compensation strategy

as evidenced by the increased harmonic distortion. This insight is crucial for understanding the operational boundaries of the CHB-VQC and for potentially developing enhancements to improve its performance across all sag severity levels.

3.5.2 *Simulation Results with Using the Proposed Optimum Control Scheme*

Illustrated in Fig. 3.9 are the graphical representations depicting the operational capabilities of the refined CHB-VQC when utilizing an optimized compensation strategy.

During the initial operational phase, the load presents an impedance of 23.325 + j20.567 Ω. When the voltage sag depth corresponds to 25% of the nominal voltage level during the interval from 0.2 to 0.3 s, the system can attain its highest output level and simultaneously minimize power energy consumption, provided the phase angle

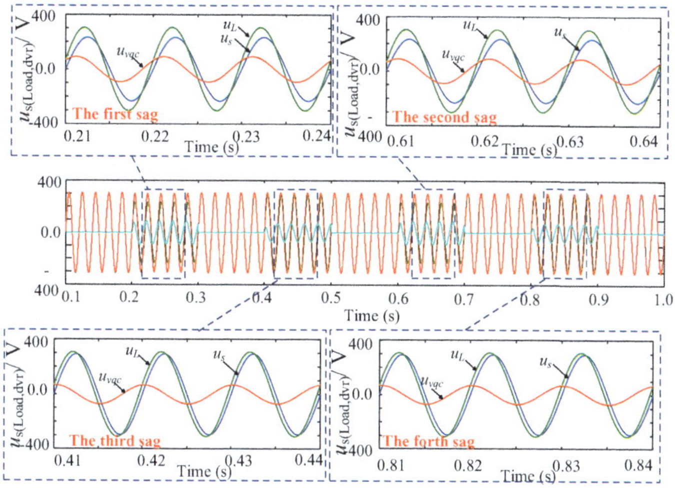

(a) source voltage u_S, compensation voltage u_{vqc} and load voltage u_{Load}

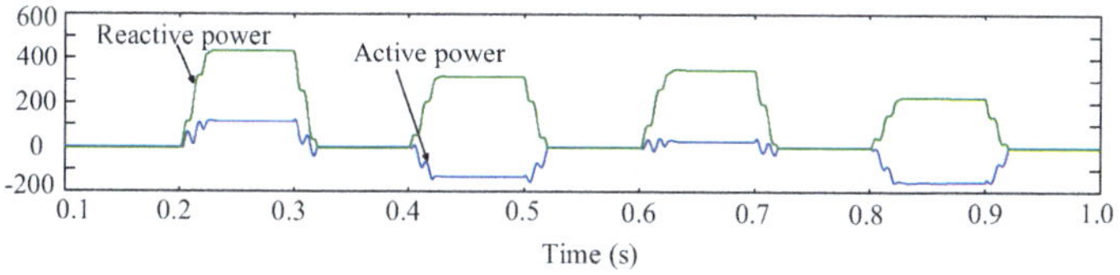

(b) active power p_{vqc} and reactive power q_{vqc}

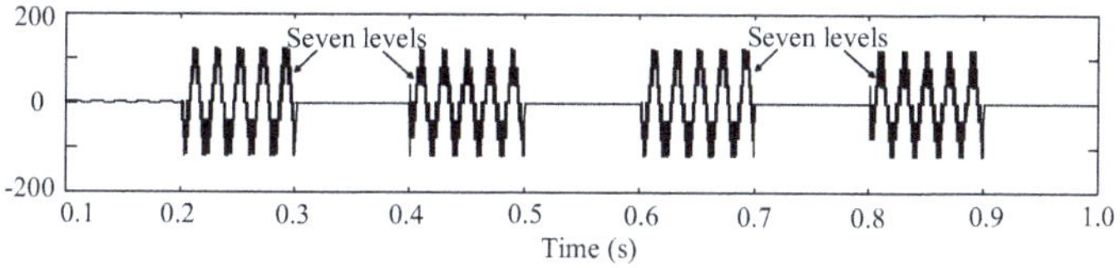

(c) stepped output voltage of the in-verter u_{IOV}

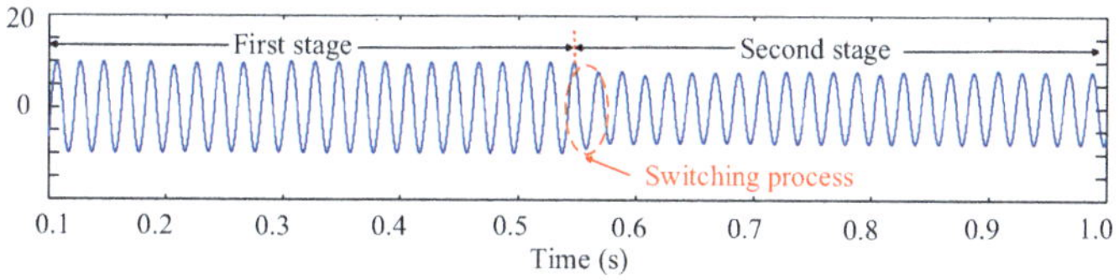

(d) load current i_{Load}

Fig. 3.9 Simulation results of the CHB-QC under in-phase compensation range of the reference voltage falls within (4.04°, 19.54°). According to Eq. (3.23), the theoretical minimum energy angle, designated as β_{min}, ought to be 41.12°, which lies outside the specified range to achieve the least active power output. Hence, the initial angle β_0 necessitates readjustment. Utilizing Eqs. (3.27) and (3.28), the recalculated β_0 is determined to be 19.54°. Subsequently, applying Eqs. (3.23) and (3.24), the magnitude and phase angle of the compensatory voltage are recalculated to generate and deliver the compensation voltage accordingly.

Simultaneously, when the voltage sag is reduced to 5% of the nominal voltage level during the period from 0.4 to 0.5 s, the phase angle range of the reference voltage is established as (14.87°, 22.64°). The theoretical minimum energy angle β_{min} in this scenario is 3.54°, which, once again, does not fall within the given phase angle range. Therefore, β_0 requires another round of adjustment, resulting in a new β_0 of 14.87°. Based on Eqs. (3.23) and (3.24), the magnitude and phase angle of the compensatory voltage are recalculated, facilitating the output of the compensation voltage.

This systematic approach ensures that the CHB-VQC operates at its most efficient state under varying sag conditions by dynamically adjusting parameters to match the specific requirements of each scenario.

In the subsequent phase, the load impedance shifts to $23.325 + j78.5\ \Omega$. When the voltage sag reaches 25% of the nominal voltage level during the time frame of 0.6 to 0.7 s, the CHB-VQC can attain its maximum output level and concurrently minimize power energy expenditure, provided the phase angle of the reference voltage is within the interval of (4.04°, 19.54°). Equation (3.23) reveals that the theoretical minimum energy angle, β_{min}, is 5.77°, which resides comfortably within the specified phase angle range. As a consequence, the CHB-VQC is able to operate at its most efficient state, achieving both minimal active power output and maximal output level.

Conversely, when the voltage sag decreases to 5% of the nominal voltage level during the duration from 0.8 to 0.9 s, the phase angle range for the reference voltage is confined to (14.87°, 22.64°). Here, the theoretical minimum energy angle, β_{min}, is calculated to be 0.90°, which is outside of the aforementioned phase angle range. In order to rectify this, Eqs. (3.27) and (3.28) are employed to adjust the initial phase angle, β_0, which is recalculated to be 14.87°. This adjustment ensures that the CHB-VQC can appropriately compensate for the voltage sag while maintaining operational efficiency under the altered conditions.

As shown in Fig. 3.9a, voltage sags occur with four different depths, the CHB-VQC can always inject a compensation voltage u_{vqc} and the load voltage u_{Load} can maintain a constant value. Figure 3.9c shows that the level number of the output voltage is always 7 when the improvement minimum energy angle scheme is adopted.

On one front, during the more pronounced sags from 0.2–0.3 s and 0.6–0.7 s, the DC link's active power output is notably reduced compared to the conventional in-phase compensation strategy. For instance, as seen in Fig. 3.9b, the DC link discharges a minimal amount of active power (approximately 100 W) during the period of 0.2–0.3 s, a stark contrast to the significant discharge (around 250 W) observed in Fig. 3.7b under similar conditions. On the opposing front, during the shallower sags at 0.4–0.5 s and 0.8–0.9 s, the DC link neither discharges nor absorbs power, instead, it facilitates the absorption of active power that is returned to the power system through the input module.

Figure 3.10 delineates the harmonic content variations for different sags when the improved minimum energy angle scheme is applied. In the initial phase depicted in Fig. 3.10a, the THD of the inverter's output voltages registers at 1.35 and 2.03% during the intervals of 0.2–0.3 s and 0.4–0.5 s, respectively. In the subsequent phase, the THDs of the inverter's output voltages are noted to be 1.54% and 2.09% in the periods of 0.6–0.7 s and 0.8–0.9 s, respectively. These figures highlight the superiority of the proposed scheme over the control strategy illustrated in Fig. 3.8, particularly for managing shallow voltage sags (occurring during 0.4–0.5 and 0.8–0.9 s).

This comparison underscores the effectiveness of the advanced minimum energy angle scheme in not only minimizing harmonic distortions but also in optimizing the power handling capacity of the CHB-VQC, especially during less severe voltage sag conditions.

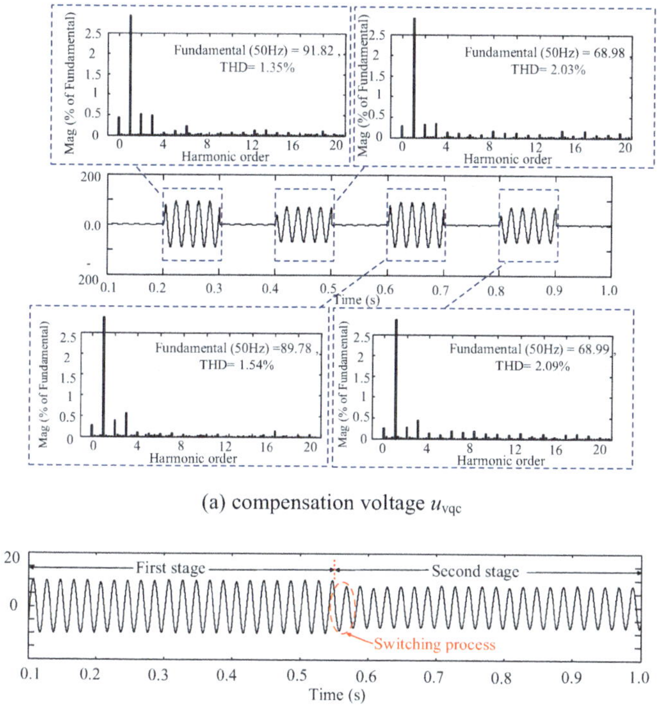

(a) compensation voltage u_{vqc}

(b) load current i_{Load}

Fig. 3.10 Harmonic content under in-phase compensation strategy

3.6 Experiment Results

A single-phase experimental verification of the proposed approach was undertaken, utilizing system parameters identical to those of the simulation, as outlined in Table 3.1. A Control Hardware-in-the-Loop (CHIL) testbed configuration was established, where the CHB-VQC was modeled in the master controller. Real-time digital simulation (RTDS) of this model was executed in the slave controller, with a time step of 10 microseconds. Control signals, sampled at 6.4 kHz, were transmitted to the I/O port of an external TMS320F2812 DSP controller via an interface card. Signal scaling factors were applied, with AC voltage and current signals scaled at 50:1 and 10:1, respectively; DC link and output stepped voltages at 40:1 and 30:1, respectively; and power at 50:1.

The performance of the CHB-VQC, employing the in-phase compensation strategy, is presented in Fig. 3.11. Specifically, Fig. 3.11a demonstrates that the source voltage experiences four sags of varying depths and durations, with depths of 25% during Δt_1 and Δt_2, and 5% during Δt_3 and Δt_4. The CHB-VQC effectively injects compensation voltage u_{vqc}, enabling the load voltage u_{Load} to remain constant. Figure 3.11b reveals that the number of output voltage levels adapts to the sag depth, with 7 levels for a 20% sag and 3 levels for a 10% sag. Additionally, the in-phase compensation strategy results in the CHB-VQC delivering both active and reactive powers during different sag scenarios, as shown in Fig. 3.7b. Lastly, Fig. 3.11c confirms the smooth transition of load current with no surges between stages, aligning with the simulation results presented in Fig. 3.7.

The performance of the CHB-VQC's enhanced minimum energy angle scheme is visually presented in Fig. 3.12. Figure 3.12a demonstrates that the source voltage experiences four distinct voltage sags at varying intervals, with depths reaching 25% of the nominal value at Δt_1 and Δt_2, and 5% at Δt_3 and Δt_4. By employing the CHB-VQC, it successfully injects compensation voltage u_{vqc}, ensuring that the load voltage u_{Load} remains stable throughout.

In Fig. 3.12b, it is evident that when the improved minimum energy angle scheme is implemented, the output voltage consistently maintains a level number of 7, outperforming the control strategy shown in Fig. 3.11b, notably during shallow voltage sags (Δt_2 and Δt_4).

Additionally, Fig. 3.12c illustrates that at Δt_1 and Δt_3, the DC link outputs reduced power compared to the traditional in-phase compensation method depicted in Fig. 3.11c. Notably, during Δt_2 and Δt_4, the DC link ceases outputting power and instead absorbs active power, which is subsequently recycled back into the power system via the input module.

Finally, Fig. 3.12d confirms that the load current transitions smoothly without any noticeable current surges between stages, aligning perfectly with the simulation results presented in Fig. 3.9. These experimental findings validate the efficacy and effectiveness of the CHB-VQC's enhanced minimum energy angle scheme.

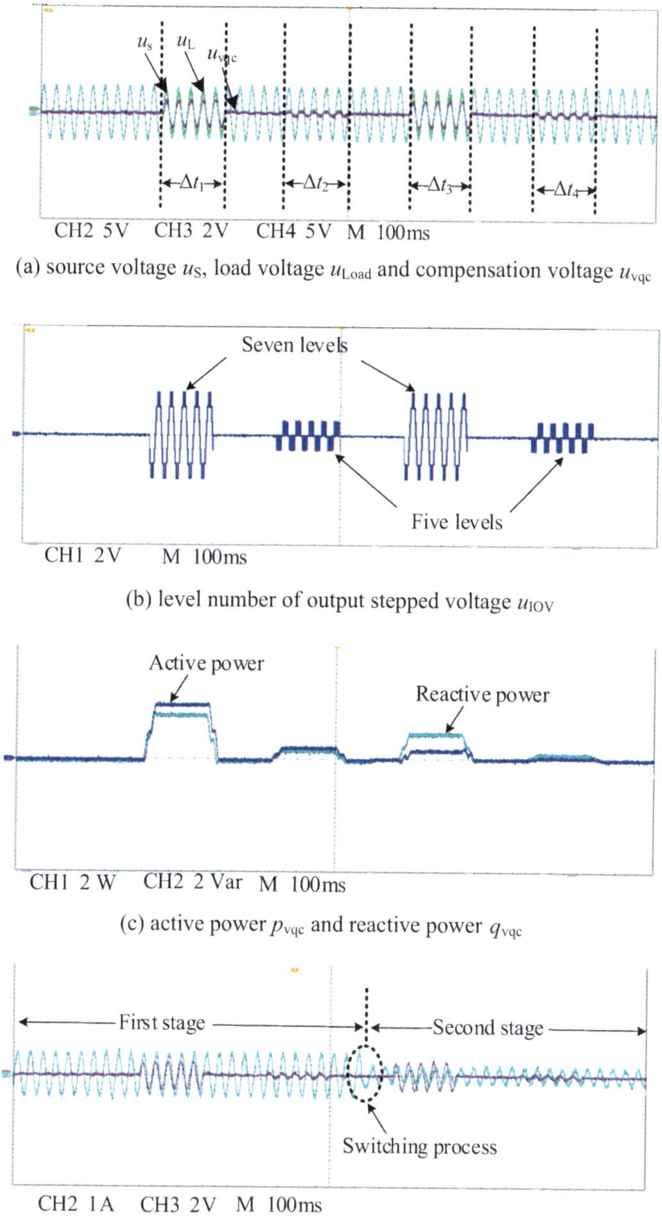

(a) source voltage u_S, load voltage u_{Load} and compensation voltage u_{vqc}

(b) level number of output stepped voltage u_{IOV}

(c) active power p_{vqc} and reactive power q_{vqc}

(d) compensation voltage u_{vqc} and load current i_L

Fig. 3.11 Experimental results of a CHB-VQC during different voltage sags

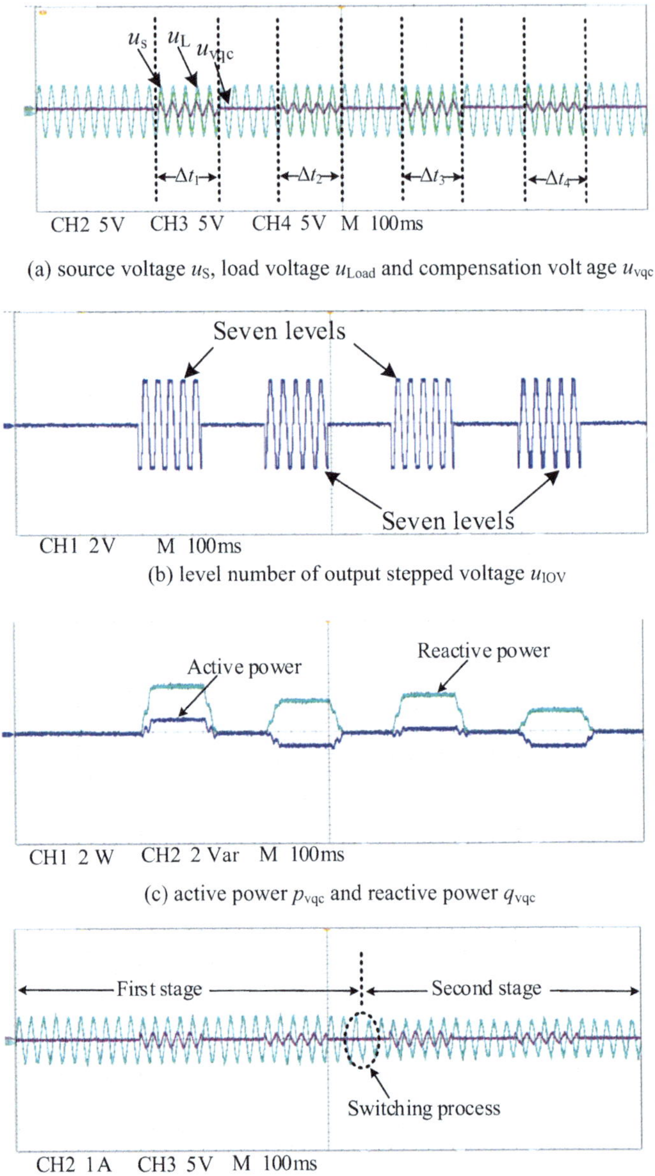

(a) source voltage u_S, load voltage u_{Load} and compensation volt age u_{vqc}

(b) level number of output stepped voltage u_{IOV}

(c) active power p_{vqc} and reactive power q_{vqc}

(d) compensation voltage u_{vqc} and load current i_L

Fig. 3.12 Experimental results of the CHB-VQC during different voltage sags

3.7 Comparative Analysis

Drawing from Figs. 3.7 through 3.10, a comparative assessment of the THD in the compensation voltages under distinct strategies is depicted in Fig. 3.13.

For Scenario I, as illustrated in Fig. 3.13a, the load impedance is characterized as 23.325 + j20.567 Ω. With a sag depth equating to 5% of the rated voltage, the THD in the compensation voltage using the proposed optimal method is remarkably lower at just 2.03%, markedly below the 6.65% observed with the conventional approach. As evidenced in Figs. 3.7 and 3.9, the output voltage level count with the proposed technique stands at 7 throughout the 5% sags, contrasting sharply with the traditional methods, which yield a mere 3 levels. Furthermore, when the sag depth escalates to 25% of the nominal voltage, the THD in the compensation voltage from the proposed strategy is virtually identical to that of the conventional method. At this juncture, both strategies produce an output voltage level count of 7, as portrayed in Figs. 3.7 and 3.9.

In Scenario II, Fig. 3.13b illustrates a shift in load impedance to 23.325 + j78.5 Ω. The THD in the compensation voltage under the proposed optimal method outperforms that of the traditional method during 5% sags, attributable to the superior number of voltage levels achieved by the former over the latter, as shown in Figs. 3.7

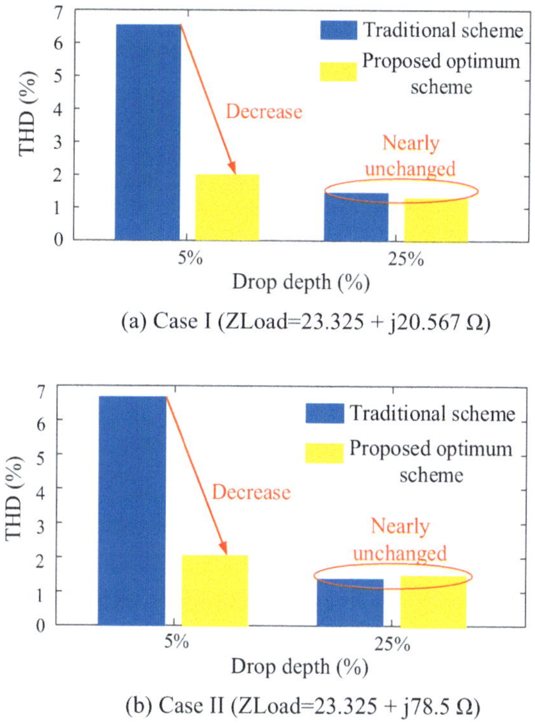

Fig. 3.13 Comparative analysis of THD of the compensation voltage during different sags

and 3.9. Moreover, during 25% sags, the THD in the compensation voltages from both methods are nearly indistinguishable due to parity in the number of voltage levels delivered.

This comparative analysis underscores the efficacy of the proposed optimal scheme in minimizing THD and enhancing voltage quality, particularly under shallow sag conditions, through the provision of a higher number of voltage levels in the compensation process.

References

1. T.V. Kiran, N. Babji, Design of a novel cost effective cascaded H-Bridge multilevel inverter topology, in *Proceeding International Conference on Signal Processing, Communication, Power and Embedded System (SCOPES), 2016, Paralakhemundi, India* (2016), pp. 726–731
2. E. Babaei, M.F. Kangarlu, M. Sabahi, Dynamic voltage restorer based on multilevel inverter with adjustable dc-link voltage. IET Power Electron. **7**(3), 576–590 (2014)
3. M. Massoud, S. Ahmed, P.N. Enjeti, B.W. Williams, Evaluation of a multilevel cascaded-type dynamic voltage restorer employing discontinuous space vector modulation. IEEE Trans. Ind. Electr. **57**(7), 2398–2410 (2010)
4. B.S. Jain, An adjustable DC link voltage-based control of multifunctional grid interfaced solar PV system. IEEE J. Ener. Select. Topic. Power Elect. **5**(2), 651–660 (2017)
5. H. Tanaka, F. Ikeda, T. Tanaka, H. Yamada, M. Okamoto, Novel reactive power control strategy based on constant DC-Capacitor voltage control for reducing the capacity of smart charger for electric vehicles on single-phase three-wire distribution feeders. IEEE J. Ener. Select. Topic. Power Elect. **4**(2), 481–488 (2016)
6. L. Li, X. Lu, H. Jiao, A balancing control method of dc side voltage for cascaded H-bridge static var generator. Trans. of China Electr. Society **31**(7), 1–7 (2016)
7. Ajami, A. Farakhor, H. Ardi, Minimisations of total harmonic distortion in cascaded transformers multilevel inverter by modifying turn ratios of the transformers and input voltage regulation. IET Power Electron. **7**(11), 2687–2694 (2014)
8. J.Y. Choi, B.M. Han. New scheme of phase-shifted carrier PWM for modular multilevel converter with redundancy sub-modules, in *Proceeding in Australasian Universities Power Engineering Conference (AUPEC), Wollongong, NSW, Australia* (2015), pp. 1–6
9. X. Xia, F. Zhang, G. Zhang, X. Jiang, Simulation and study on cascaded nineteen-level dynamic voltage restorer, in *Proceeding in Asia-Pacific Power and Energy Engineering Conference, Shanghai, China* (2012), pp. 1–4
10. X. Guo, D. Xu, J.M. Guerrero, B. Wu, Space vector modulation for DC-Link current ripple reduction in back-to-back current-source converters for microgrid applications. IEEE Trans. Ind. Electron. **62**(10), 6008–6013 (2015)
11. M. Rauf, V. Khadkikar, An enhanced voltage sag compensation scheme for dynamic voltage restorer. IEEE Trans. Indus. Electron. **62**(5), 2683–2692 (2015)
12. R.D. Meyer, Y. Li, F. Blaabjerg, Optimized control strategy for a medium-moltage DVR-theoretical investigations and experimental results. IEEE Trans. Power Electron. **23**(6), 2746–2754 (2008)
13. Z. Sun, C. Guo, X. Xiang, Analysis method of DVR compensation strategy based on load voltage and minimum energy control. Proc. CSEE **30**(31), 43–49 (2010)
14. B.F. Citarsa, I.N.W. Satiawan, Supriono, a new modulation technique for a three-cell single-phase CHB inverter with un-equal DC-link voltage for improving output voltage quality, in *Proceeding in 2nd International Conference on Applied Electromagnetic Technology (AEMT), Lombok, Indonesia* (2018), pp. 73–78

15. A. Sadaba, P.S. Gurpide, J.L. Taberna, I.M. Morales, L.M. Palomo, Voltage harmonics generated by 3-level converters using PWM natural sampling, in *Proceeding in IEEE 32nd Annual Power Electronics Specialists Conference, Vancouver, BC, Canada* (2001), pp. 1565–1561
16. X. Ma, F. Wang, W. Blaabjerg, S. Song, T.L. Wang, Real-time calculation method for single-phase cascaded H-bridge inverters based on phase-shifted carrier pulse width modulation. IEEE Trans. Power Electron. (2019). https://doi.org/10.1109/TPEL.2019.2911422
17. Ilves, L. Harnefors, S. Norrga, H.P. Nee, Analysis and operation of modular multilevel converters with phase-shifted carrier PWM. IEEE Trans. Power Electron. **31**(1), 268–283 (2015)
18. Y. Li, Y. Wang, B.Q. Li, Generalized theory of phase-shifted carrier PWM for cascaded H-bridge converters and modular multilevel converters. IEEE J. Emerg. Select. Topics Power Electron. **4**(2), 589–605 (2016)
19. B. Li, R. Wang, D. Xu, G. Wang, W. Wang, D. Xu, Analysis of the phase-shifted carrier modulation for modular multilevel converters. IEEE Trans. Power Electron. **30**(1), 297–310 (2015)

Part II
Dual-Functional Voltage Quality Controller

By proposing a new dual-functional voltage quality controller (DF-VQC) concept, VQC equipment utilization is improved. Building new topologies and having them perform different functions in different modes. The first type of DF-VQC uses a crowbar bidirectional thyristor switch across the output terminals of a conventional back-to-back VQC. In the event of load short, the VQC controller will deactivate the faulty phase of the VQC and activate its crowbar thyristor to insert the VQC filter reactor into the grid to limit the fault current. The second type of DF-VQC is mainly to set up a brake chopper circuit and a crowbar bidirectional thyristor separately on the inverter side. When the power grid voltage drops, the crowbar bidirectional thyristor switches off to protect the left rectifier unit, while the right inverter side and the brake chopper circuit form a bridge current limiter to achieve fault current limiting function.

Chapter 4
Dual-functional Voltage Quality Controller Based on Filter Inductor Multiplexing

This section proposes a new DF-VQC concept. The new topology uses a crowbar bidirectional thyristor switch across the output terminals of a conventional back-to-back VQC. The design methodology of the proposed FCL-VQC is fully discussed based on the PSCAD/EMTDC simulation.

4.1 Introduction

In this book, a new concept of DF-VQC is proposed. The new topology can operate in two operational modes: (1) compensation mode for voltage fluctuation and unbalance, and (2) fault current-limiting mode. It should be noted that compared with a conversational VQC, the new DF-VQC is designed by adding an additional crowbar bidirectional thyristor switch across the output terminals of each phase of the conventional VQC. Thus, the power rating of DF-VQC will be the same as the conventional VQC without FCL function. This section is structured as follows. In Sect. 4.2, the topology, principle, and the control strategy are proposed. In Sect. 4.3, the design methodology of important system parameters of DF-VQC is discussed. In Sect. 4.4, the effectiveness of DF-VQC is verified by the simulation and the experimental results.

F. Jiang et al., *Advanced Voltage Quality Controller for New Power Distribution Systems*, https://doi.org/10.1007/978-981-97-9861-2_4

4.2 Topology and Principle of Operation

4.2.1 Topology

The topology of the DF-VQC is shown in Fig. 4.1. It is composed of three single-phase bridges. Each single-phase topology mainly comprises a shunt transformer, a back-to-back power converter, a series transformer and a crowbar bidirectional thyristor. The input rectifier module of the back-to-back converter is connected to the grid through a shunt transformer with L_Z to eliminate the high-frequency ripples, and rectifies the power from the grid to the DC link capacitor. The output inverter module converts the power from the DC link capacitor to compensate voltage fluctuation, and is connected to the grid through a series transformer and a LC output filter (e.g., L_a, C_a). The PWM rectifier and inverter are connected through the DC link capacitor C_d. The crowbar bidirectional thyristor in each phase (e.g., SCR_a in Phase-A) is across the output terminals of the PWM inverter to provide short-circuit fault current-limiting function. u_s, u_{dc}, u_c, u_{vqc}, and u_l represent the supply voltage, the DC link voltage, the output voltage of the DF-VQC, the output voltage of DF-VQC on the primary side of the series transformer, and the voltage of point of common coupling (PCC), respectively. i_s and i_l are the supply current and the load current, respectively. Z_s and Z_l are the equivalent impedance of the grid and the equivalent impedance of the transmission line, respectively. In this book, the subscripts a, b, and c mean phase-A, phase-B, and phase-C, respectively.

As shown in Fig. 4.1, the major difference between the new DF-VQC and a conventional VQC is the addition of three crowbar bidirectional thyristors. The crowbar bidirectional thyristor for each phase will be deactivated or activated depending on

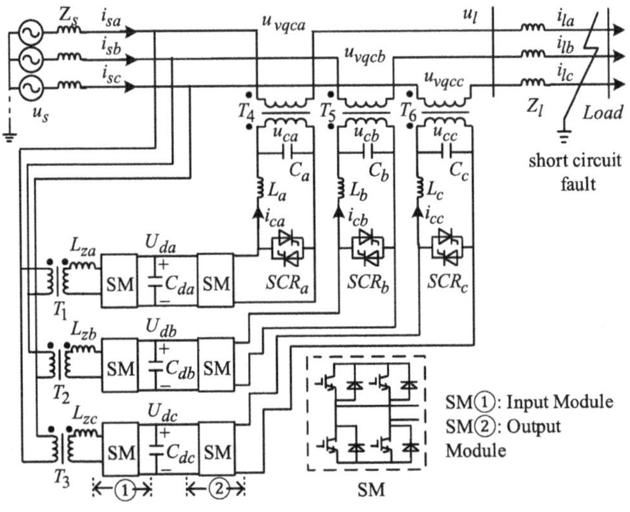

Fig. 4.1 Topology of DF-VQC

whether a fault condition is detected or not. When the grid is under normal operation state, the crowbar bidirectional thyristor is deactivated and the DF-VQC operates in the voltage compensation mode to compensate voltage fluctuations. When a short-circuit fault occurs, the crowbar bidirectional thyristor is activated to insert the output inductor into the main current path through the series transformer. At the same time, the IGBTs of the PWM inverter will be turned off to completely deactivate the inverter. The DF-VQC thus operates in the fault current-limiting mode. The short-circuit fault current can be limited by L through the crowbar bidirectional thyristor. The maximum fault current can also be controlled by adjusting the conduction angle of the thyristor. Therefore, both voltage compensation and fault current-limiting functions can be provided by the DF-VQC with the same power rating of a conventional VQC.

When a short-circuit fault or a voltage sag situation occurs, the fluctuation of supply voltage will influence the stability of DC link voltage. If the three shunt transformers are delta-connected on the high-voltage side, the voltage fluctuation on the low-voltage side is smaller than the Y-connection mode. Meanwhile, delta-connection can also suppress the 3rd harmonics caused by the fluctuation of DC link voltage. So in this book, the high-voltage sides of the shunt transformers are delta-connected.

4.2.2 Principle of Operation

(1) Voltage Compensation Mode

As shown in Fig. 4.2, when the voltage compensation mode is applied, DF-VQC is equivalent to a voltage control voltage source. When the voltage in the system fluctuates, the voltage of DF-VQC can be seen as the compensation voltage U_1, which can be connected in series with the power supply voltage to maintain the load voltage. This not only improves the power quality, but also achieves the same effect.

(2) Fault Current-Limiting Mode

A single-phase equivalent circuit showed in Fig. 4.3 used to achieve current-limiting mode in the event of a fault. When the system encounters a short-circuit fault, the faulty phase in the inverter will be automatically disabled, and at the same time, the crowbar bidirectional thyristor will be activated to cope with the fault situation. This design can quickly take measures in case of circuit abnormalities, ensuring the

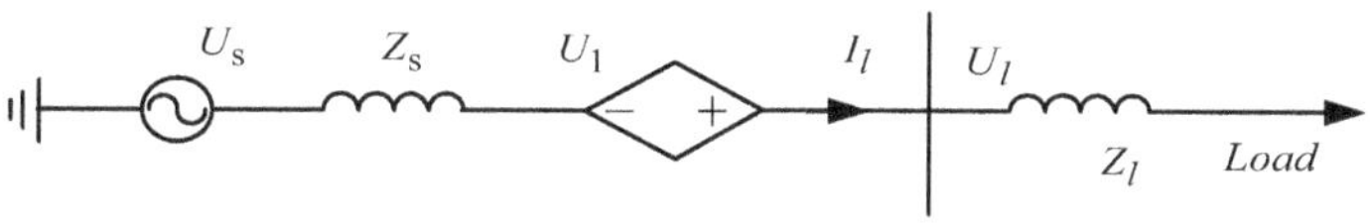

Fig. 4.2 Equivalent circuit of DF-VQC when it operates in voltage compensation mode

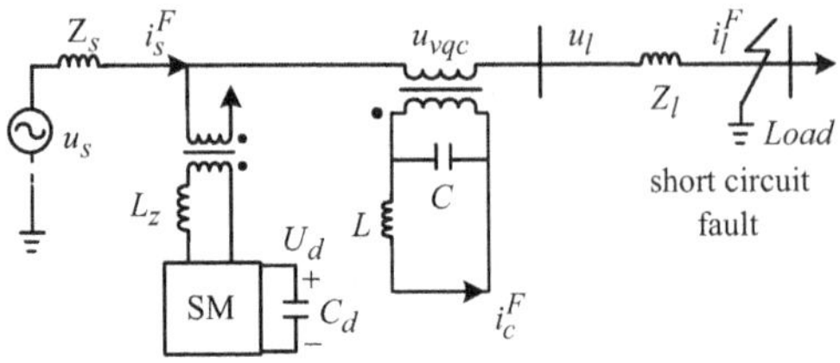

Fig. 4.3 Single-phase equivalent circuit of DF-VQC when it operates in fault current-limiting mode

stability and safety of the system. In fault current-limiting mode, the circuit adopts a series of measures to effectively address the problem. Especially, reactor L will be introduced into the power grid on the secondary side of the series transformer. The purpose of doing so is to limit the flow of fault current, prevent it from causing more serious damage to the system, and thus ensure the normal operation and safe operation of the equipment. Furthermore, it should be emphasized that series transformers play a crucial role in this process. Because the power supply voltage is almost entirely maintained by series transformers, this means that even in the event of a fault, the system can still maintain a certain degree of stability. This provides an important foundation and guarantee for subsequent fault recovery. In actual operation, the input rectifier module (PWM rectifier) of the converter will maintain normal operation. This is crucial as it ensures that the system can more easily recover and repair after a failure occurs.

In Fig. 4.3, the strategy for suppressing high-frequency fluctuations involves resonance of the L and C at relatively high frequencies. The resonance frequencies can be determined by equation $n^2\omega_0 L = 1/\omega_0 C$. Where ω_0 represents the fundamental frequency, and n usually exceeds 10. Because the equivalent impedance of L is significantly smaller than that of C, the influence of capacitance during the filtering process can be ignored. The following formula can be used to calculate:

$$Z_{\mathrm{eq}} = k^2\omega_0 L. \tag{4.1}$$

The equivalent impedance of L on the primary side is

$$Z_{\mathrm{eq}} = k^2\omega_0 L, \tag{4.2}$$

where k is the ratio of series transformer.

When the DF-VQC operates in the fault current-limiting mode, the supply current is given by

$$I_s^F \approx I_c^F / k \approx I_l^F \approx \frac{U_s}{Z_s + Z_{\mathrm{eq}} + Z_l}, \tag{4.3}$$

from (4.3), it is evident that the output reactor can restrict the short-circuit current. Given that Z_s and Z_l are typically smaller than Z_{eq}, fault current is predominantly governed by Z_{eq}. The parameters of k and L, along with the adjustment of the conduction angle of the crowbar thyristor, can constrain the fault current to the specified value.

4.2.3 Control Strategy of DF-VQC

Displayed in Fig. 4.4 is the comprehensive control block diagram of the DF-VQC. Within this diagram, Fig. 4.4a delineates the intricate control mechanism of the PWM rectifier, meticulously designed to uphold the stability of the DC link voltage. Conversely, Fig. 4.4b provides insight into the control block diagram governing the inverter and the bidirectional thyristor crowbar, pivotal in executing essential functions such as voltage compensation and fault current limitation.

In consonance with prior discussions, the operational efficacy of the DF-VQC hinges significantly upon two pivotal factors: the strategic switching between operational modes and the efficacy of the fault current detection method. In accordance with grid conditions, the DF-VQC seamlessly transitions between two operational modes. While in the voltage compensation mode, it adeptly functions as a conversational VQC, aligning itself with established control methodologies outlined in references [1–4]. However, the focus of this book primarily revolves around delineating the intricacies of the switching strategy and the fault current detection method, particularly concerning the fault current-limiting operational mode.

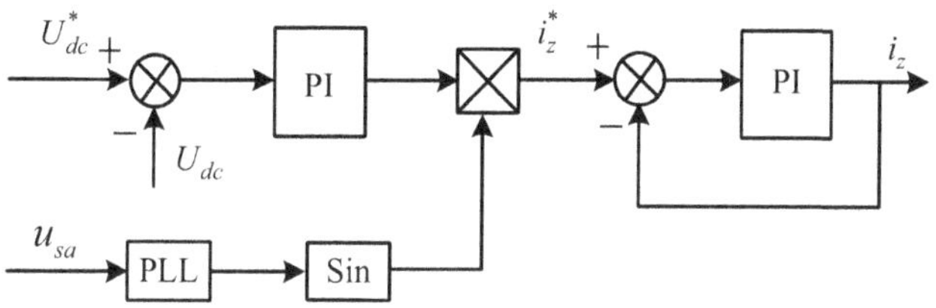

(a) Control block diagram of rectifier to maintain the DC link voltage

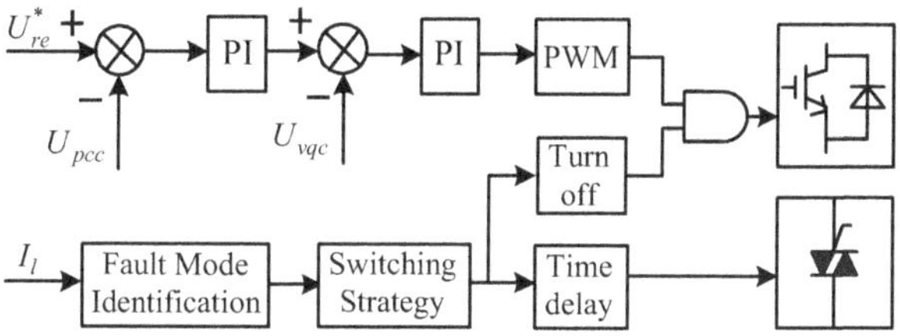

(b) Control block diagram of the inverter and crowbar bidirectional thyristor

Fig. 4.4 Control block diagram of the DF-VQC

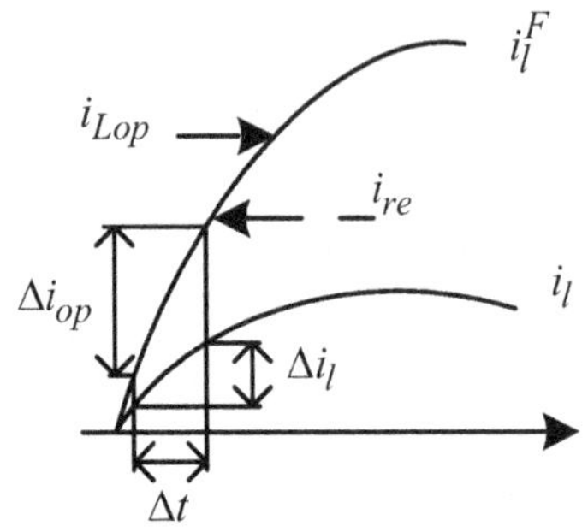

Fig. 4.5 Fault current detection method by sensing load current and its rate of change

Within Fig. 4.4a, the book embraces a control methodology grounded in the instantaneous values of critical parameters, including the DC link voltage, input current of the PWM rectifier, and the Point of Common Coupling (PCC) voltage, as expounded in reference [4]. By leveraging this meticulously designed control approach, the DF-VQC ensures swift response times and mitigates steady-state errors inherent in voltage compensation mechanisms. Furthermore, the three-phase topology, comprised of three single-phase bridges, affords granular control over each phase, facilitating the seamless implementation of unbalance compensation strategies to address grid asymmetries and enhance overall system performance.

If the fault current detection is too slow, the stability of the grid will be influenced. On the other hand, if the fault current detection is inaccurate, the short-circuit current-limiting function of the DF-VQC will not be performed properly. Thus a fault current detection method by sensing the load current and its rate of change is developed, as shown in Fig. 4.5.

In the proposed method of fault current detection, if the following two conditions can be both met, short-circuit fault can be judged. (a) For the k_1 consecutive sampling period, the instantaneous value of load current i_{LOP} is larger than the reference value i_{re}; (b) For the k_2 consecutive sampling period, the difference between the adjacent sampling values of load current Δi_L is larger than the reference value Δi_{OP}. In this book, $k_1 = (4 \sim 8)$, and $k_2 = (4 \sim 8)$.

The operation mode switching strategy is summarized as the following:

Case 1: The grid is in the normal state, and no short-circuit fault occurs. DF-VQC operates in voltage compensation mode; and the crowbar bidirectional thyristors are deactivated. In this state, the compensation voltage of DF-VQC is zero. It acts as a pass-through for the active transmission.

Case 2: PCC voltage fluctuations/imbalances and no faults occur. DF-VQC operates in voltage compensation mode and the crowbar bidirectional thyristors remains the same status as in case 1.

Case 3: Transient single phase to ground fault occurs. DF-VQC operates in voltage compensation mode, and the crowbar bidirectional thyristors are deactivated. DF-VQC compensates the PCC voltage. The power grid resumes to a normal state automatically while the fault disappears.

Case 4: Permanent single phase to ground fault occurs. The healthy phase of DF-VQC operates in voltage compensation mode, and the corresponding crowbar bidirectional thyristors are deactivated. The fault phases of DF-VQC are out of operation, and will recovery to voltage compensation mode while the permanent single phase to ground fault disappears.

Case 5: Transient short-circuit faults (i.e., phase-to-phase short-circuit fault, two-phase to ground fault and three phase to ground fault) occur. The healthy phase of DF-VQC operates in voltage compensation mode; the fault phases are in fault current-limiting mode. The voltage compensation command of the fault phase is set to zero, and the inverter of the fault phase is deactivated. The corresponding crowbar bidirectional thyristors are activated. The PWM rectifier of the fault phase operates well to maintain the DC voltage, and it prepares for the recovery of DF-VQC. When the fault disappears, the grid resumes to normal operation status automatically; and the fault phase of DF-VQC recoveries to voltage compensation mode.

Case 6: Permanent short-circuit faults (i.e., phase-to-phase short-circuit fault, two-phase to ground fault, and three phase to ground fault) occur. Like in case 5, the healthy phase of DF-VQC operates in voltage compensation mode; the fault phases are in the fault current-limiting mode. The fault phase of back-to-back converter is deactivated, and the corresponding crowbar bidirectional thyristors are activated. Until the faults are eliminated, the fault phase of DF-VQC reworks.

4.3 Optimal Selection of DF-VQC Parameters

In this section, the coupling relationship between the main parameters and the compensation capacity is analyzed, and an optimal parameter selecting method of main circuit parameter is proposed. The flowchart of parameters selecting method is shown in Fig. 4.6. According to the state of the grid, the compensation capacity of DF-VQC and the value of DC link voltage are designed firstly. Then other parameters of DF-VQC are designed according to the DC link voltage and compensation capacity. As a result of linkages between some parameters, the DC link voltage has played a role of bridge. When the parameters are designed, the DC link voltage should be recalculated until all parts of DF-VQC meet the requirement.

4.3.1 Compensation Capacity Design and DC-Side Voltage Selection

4.3.1.1 Compensation Capacity Design

Assuming the rated apparent capacity of load is S_{LOAD}, DF-VQC has a compensation ability to keep the load voltage stability when voltage fluctuation is up to $m\%$. $\cos\varphi$ is

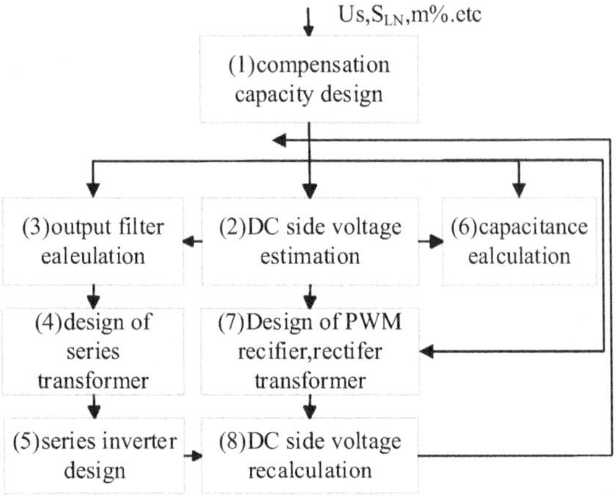

Fig. 4.6 Flowchart of optimal parameter selecting method

the power factor of the load, and it doesn't change before and after voltage fluctuation occur. The equivalent resistor of the load is

$$R_{\text{eq}} = \frac{U_s^2}{(\cos\varphi S_{\text{LOAD}})/3} * (\cos\varphi)^2 = \frac{3U_s^2}{S_{\text{LOAD}}} * \cos\varphi. \tag{4.4}$$

Then the active power of the load before and after voltage fluctuation occurs is

$$\begin{cases} p_1 = \dfrac{\cos\varphi S_{\text{LOAD}}}{3} \\ p_2 = \dfrac{U_s^2(1 \pm m\%)^2}{R_{\text{eq}}} * (\cos\varphi)^2 = \dfrac{\cos\varphi S_{\text{LOAD}}}{3}(1 \pm m\%)^2 \end{cases}, \tag{4.5}$$

where + indicates voltage swell, – indicates voltage sag.

So the active power that the DF-VQC should be compensated for voltage fluctuation should be

$$\Delta p = p_1 - p_2 = \frac{\cos\varphi S_{\text{LOAD}}}{3}[1 - (1 \pm m\%)^2], \tag{4.6}$$

Taking into account the loss P_{loss} of the series transformer, PWM rectifier and so on, the active power provided by the PWM rectifier should be

$$P_D > \Delta p + P_{\text{loss}}. \tag{4.7}$$

At the same time, P_D must be required to leave some margin in order to meet the compensation in case of overload.

4.3.1.2 DC-Side Voltage Selection

The DC link voltage should be higher than both the secondary side peak voltage of series transformer and shunt transformer. As the ratio of the series transformer and shunt transformer is not determined, according to the collector-emitter voltage of the IGBT, the DC link voltage could be estimated as

$$U_{\mathrm{dc}} \leq 0.65 V_{\mathrm{CES}}, \tag{4.8}$$

where V_{CES} is the collector-emitter voltage of IGBT.

4.3.2 *Design of LC Output Filter, Series Transformer, and PWM Inverter*

4.3.2.1 Series Transformer Design

Setting the resonance frequency of LC filter is f_c, then $2\pi f_c L = 1/(2\pi f_c C)$. To filter out the switching ripples, harmonics voltage should try to drop on the reactor, leave the output compensation voltage land on the capacitor.

The inverter could be equivalent to an equivalent resistance

$$R_L = \frac{U_{\mathrm{dc}}^2}{P_D}. \tag{4.9}$$

The resonance frequency is

$$f_c = \frac{1}{2\pi\sqrt{\mathrm{LC}}} = \frac{\sqrt{L/C}}{2\pi L}. \tag{4.10}$$

Assuming $\rho = (L/C)^{1/2}$, typically in engineering, usually there is $\rho = (0.5 \sim 0.8)R_L$. Then according to (4.9) and (4.10), L and C can be calculated as follows

$$\begin{cases} L = \dfrac{\rho}{2\pi f_c} \\ C = \dfrac{L}{\rho^2} = \dfrac{1}{2\pi f_c \rho} \end{cases}, \tag{4.11}$$

the rated current of L is not only determined by the capacity of the load but also by the value of L, u_s, Z_s, Z_l, and k.

4.3.2.2 Series Transformer Design

The rated load current is

$$I_{\mathrm{Lmax}} = \frac{S_{\mathrm{LOAD}}}{3U_s}. \tag{4.12}$$

The ratio of the series transformer is in connected with the target of fault current-limiting result. When three-phase short-circuit occurs, the fault current is the maximum. Assuming that the maximum fault current isλ times of the rated load current

$$I_{\mathrm{sF}} \approx \lambda I_{\mathrm{Lmax}}. \tag{4.13}$$

To meet the requirement of limiting, the equivalent impedance of output reactor L in primary side should be

$$Z_{\mathrm{eq}} = \sqrt{\left(\frac{U_{\mathrm{s}}}{\lambda I_{\mathrm{Lmax}}}\right)^2 - Z_s^2 - Z_l^2}. \tag{4.14}$$

According to (4.11) and (4.14), the ratio of the series transformer can be calculated

$$k = \frac{\left[\left(\frac{U_{\mathrm{s}}}{\lambda I_{\mathrm{Lmax}}}\right)^2 - Z_s^2 - Z_l^2\right]^{\frac{1}{4}}}{\sqrt{\omega_0 L}}. \tag{4.15}$$

Compared with Z_{eq}, Z_s and Z_l are small enough to be negligible, the ratio of the series transformer should be

$$k \approx \sqrt{\frac{U_{\mathrm{s}}}{\lambda I_{\mathrm{Lmax}} \omega_0 L}}. \tag{4.16}$$

When the DF-VQC operates in fault current-limiting mode, the primary side of series transformer may withstand the supply voltage, so the capacity of series transformer is

$$S_T = U_{\mathrm{s}} I_s^F = \lambda U_{\mathrm{s}} I_{\mathrm{Lmax}}. \tag{4.17}$$

Also a certain margin must be taken into considering. In order to reduce the loss of the transformer, the equivalent impedance of the transformer should be designed as small as possible.

4.3.2.3 PWM Inverter Design

Assuming the DF-VQC has a compensation ability to keep the load voltage stability when supply voltage fluctuation up to $m\%$, then the output voltage of the inverter should be

$$U_c^* = \frac{mU_s}{100k}. \tag{4.18}$$

The equivalent circuit of the series PWM inverter is shown in Fig. 4.7. In Fig. 4.7, the output voltage of the inverter is

$$U_o = jk\omega_0 \mathrm{LI}_L + U_c. \tag{4.19}$$

When the grid runs in a state of maximum load, voltage drop on the filter inductor of series inverter is

$$U_{\mathrm{Lp}} = jkw_0\mathrm{LI}_{L\max} = \frac{jkw_0\mathrm{LS}_{\mathrm{LOAD}}}{3U_s}. \tag{4.20}$$

When the grid operates in a state of rated load and compensation voltage of series inverter U_c is in phase with the load current I_L, and the voltage drop U_{Ld} on the filter reactor leads the load current 90°, the active power of series inverter reaches the maximum. In this state, the output fundamental voltage U_o reaches the maximum value of $U_{\mathrm{omax}} = m_2 U_{\mathrm{dc}}/2^{1/2}$, where m_2 is modulation ratio of the inverter, then

$$U_c = \sqrt{U_{\mathrm{omax}}^2 - U_{\mathrm{Ld}}^2}, \tag{4.21}$$

Therefore the theoretical rated compensation active power capacity of DF-VQC can be calculated as

$$P_T = kI_{\mathrm{Lmax}}\sqrt{U_{\mathrm{omax}}^2 - U_{\mathrm{Ld}}^2}, \tag{4.22}$$

The total capacity of the series inverter S_D is

$$S_D = kU_{\mathrm{omax}}I_{\mathrm{Lmax}}. \tag{4.23}$$

Fig. 4.7 Equivalent circuit of the DAB output

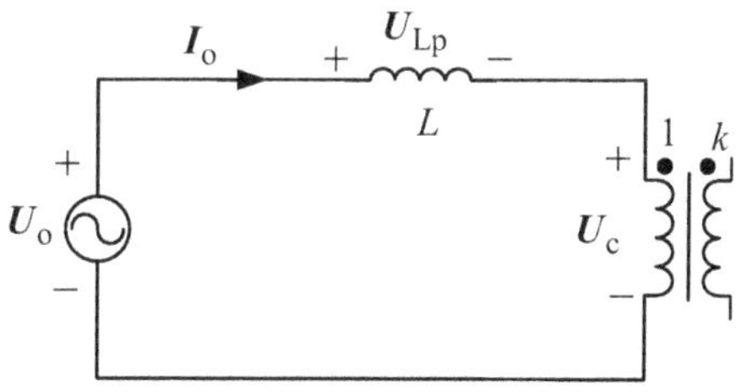

When short-circuit fault occurs, before the current-limiting operation is completed, fault current always increases and flow through the inverters until all IGBTs are deactivated, so certain margin of the capacity must be considered.

4.3.3 DC Link Capacitor Selection and Shunt Transformer Design

4.3.3.1 DC Link Capacitor Selection

More capacitance cuts DC fluctuation but boosts controller inertia, hampering real-time control. To stabilize DC voltage, rectifier active power should exceed output slightly, so DC capacitor calculation considers (P_D). Assuming max allowable ripple (ρ_v), compute DC capacitance with Eq. (4.24).

$$C_d \geq \frac{P_D}{2\omega_0 \sigma_V U_{dc}^2}. \tag{4.24}$$

4.3.3.2 PWM Rectifier and Shunt Transformer Design

The operation phasor diagram of the PWM rectifier is showed in Fig. 4.8, where U_{sz}, U_{LZ}, U_{OZ}, I_Z are the secondary voltage of the shunt transformer, the voltage drop on the filter reactor I_Z, output voltage and output current of the rectifier, respectively. From Fig. 4.8, the following relationship can be obtained

$$U_{SZ} = U_{LZ} + U_{OZ} = j\omega_0 L_Z I_Z + U_{OZ}. \tag{4.25}$$

The AC side current of the rectifier I_Z is in phase with or reverse with supply voltage U_{SZ}, then the phase angle of voltage drop on filter reactor U_{LZ} leads or lags the supply voltage U_{SZ} is 90^0, so the following relationship can be derived.

$$U_{oz}^2 = U_{sz}^2 + U_{Lz}^2 = U_{sz}^2 + (\omega_0 L_z I_z)^2, \tag{4.26}$$

so I_Z can be calculated as in (4.27)

$$I_z = \frac{\sqrt{U_{oz}^2 - U_{sz}^2}}{\omega_0 L_z}, \tag{4.27}$$

assuming the transformer ratio is n:1, then $U_{sz} = U_{sl}/n$. The maximum output current of the rectifier I_{zmax} is

Fig. 4.8 Operation phasor diagram of PWM rectifier

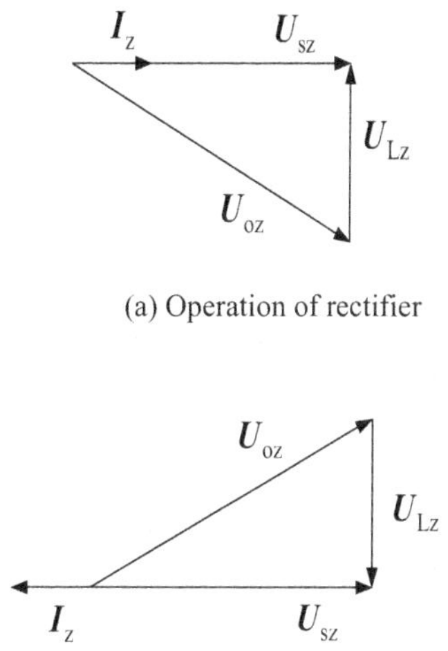

$$I_{zmax} = \frac{\sqrt{U_{ozm}^2 - (U_{sl}/n)^2}}{\omega_0 L_z}. \tag{4.28}$$

When the line voltage drop $m\%$, the PWM rectifier will absorb active power from the grid to maintain the stability of DC link voltage. The output current of the rectifier I_z becomes large because of U_{sz} becomes small, in order to reduce the over current capability of rectifier bridge, the current of rectifier should be less than I_{lim} when line voltage drop $m\%$

$$I_z = \frac{nP_D}{U_{sl}(1 - m\%)} < I_{lim}. \tag{4.29}$$

The line voltage will rise of $m\%$ when phase voltage rise of $m\%$, the maximum output voltage of series converter is $U_{ozm} = m_1 U_{dc}/2^{1/2}$, where m_1 is modulation ratio of PWM rectifier, and in this condition U_{SZ} may be greater than U_{OZM}. When U_{sl} rise of $m\%$, the rectifier transformer ratio must meet the following expression

$$U_{ozm} > \frac{U_{sl}}{n}(1 + m\%). \tag{4.30}$$

According to (4.29) and (4.30), the range of the transformer ratio is

$$\frac{\sqrt{2}U_{sl}(1 + m\%)}{m_1 U_{dc}} < n < \frac{I_{lim} U_{sl}(1 - m\%)}{P_D}. \tag{4.31}$$

Assuming the ripple current factor is σ, the relationship of ripple current ΔI_Z, instantaneous output voltage u_{OZ}, the duty cycle $D(t)$, the switching frequency f_k is

$$\Delta I_z = \frac{U_{dc} - u_{oz}(t)}{L_z} \cdot \frac{D(t)}{f_k}, \tag{4.32}$$

when $U_{oz}(t) = U_{dc}/2$ and other parameters keep invariant, the maximum ripple current of I_Z is

$$\Delta I_{zmax} = \frac{U_{dc}}{4L_z f_k}. \tag{4.33}$$

The filter reactor I_Z should be

$$L_z \geq \frac{U_{dc}}{4f_k \Delta I_{zmax}} = \frac{U_{dc}}{4f_k \sigma_i I_{zmax}}, \tag{4.34}$$

putting (4.28) into (4.34)

$$n \geq \frac{U_{sl}}{\sqrt{U_{ozm}^2 - \left(\frac{\omega_0 U_{dc}}{4\sigma_i f_c}\right)^2}}. \tag{4.35}$$

The ranges of rectifier transformer ratio can be determined by (4.31) and (4.35).

The active power capacity of the rectifier is $P_D = U_{sz} I_{zmax}$, when U_{oz} is the maximum, namely $U_{ozm} = m_1 U_{dc}/2^{1/2}$, the current of PWM rectifier reach the maximum I_{zmax}, with (4.28) we can obtain

$$\frac{U_{sl}}{n} \cdot \frac{\sqrt{U_{ozm}^2 - (U_{sl}/n)^2}}{\omega_0 L_z} \geq P_D, \tag{4.36}$$

putting n into (4.36), L_Z should be

$$\begin{cases} L_z \leq \dfrac{U_{sl}(1+m\%)\sqrt{U_{ozm}^2 - (U_{sl}(1+m\%)/n)^2}}{n\omega_0 P_D} \\ L_z \leq \dfrac{U_{sl}(1-m\%)\sqrt{U_{ozm}^2 - (U_{sl}(1-m\%)/n)^2}}{n\omega_0 P_D} \end{cases}. \tag{4.37}$$

4.3.4 Recalculation of the DC-Side Voltage

Since the initial DC link voltage is estimated, and the DC link voltage connects PWM rectifier with the inverter. So the estimated DC link voltage could meet the requirements of each state. According to (4.21) and the capacity of the inverter, the output voltage of the inverter should be

$$U_c > \frac{mU_s}{100k}. \tag{4.38}$$

The DC link voltage should meet

$$\sqrt{\left(\frac{m_2 U_{dc}}{\sqrt{2}}\right)^2 - \left(\frac{k\omega_0 \mathrm{LS}_{\mathrm{LOAD}}}{3U_s}\right)^2} > \frac{mU_s}{100}. \tag{4.39}$$

According to (4.30) and the capacity of rectifier, the DC link voltage should subject to (4.40)

$$\frac{m_1 U_{dc}}{\sqrt{2}} > \frac{U_{s1}(1 + m\%)}{n}. \tag{4.40}$$

If the DC link voltage does not satisfy (4.39) and (4.40), the DC link voltage and parameters should be recalculated.

4.4 Simulation and Experimental Results

4.4.1 Simulation Case Study

PSCAD/EMTDC simulation is carried out to verify the validity of the proposed topology and design methodology. The supply voltage is set at 10 kV with a 1 MW resistive load. The DF-VQC is designed to compensate a voltage fluctuation of 20% of the supply voltage. The maximum short-circuit fault current is allowed to be 6 times of the nominal load current. The parameters are summarized in the Table 4.1.

Table 4.1 Parameters of simulation

Parameters	Value
Compensation capacity	132 KVA
DC link voltage	800 V
LC filter of series converter	0.25 mH, 27 uF
Series transformer ratio	8:1
DC link capacitance	15,000 uF
Shunt transformer ratio	23.5
Output reactor of PWM rectifier	1 mH

4.4.2 Voltage Compensation Function and Fault Current-Limiting Function of DF-VQC

4.4.2.1 Voltage Compensation Function of DF-VQC

Figure 4.9a and b show the voltage compensation performance of the DF-VQC for voltage fluctuation and unbalance. Us, U_L, I_L, U_{dc}, and U_{VQC} are the supply voltage, load voltage at the point of common coupling, the load current, the DC link voltage, and the output voltage of DF-VQC in the primary side of inverter.

In Fig. 4.9a, voltage sag happens between 0.5 and 0.6 s with a depth of 20%. When the DF-VQC is put into operation at 0.5 s, the inverter of the DF-VQC output a compensation voltage U_{VQC} within one cycle (of 50 Hz) by absorbing active power from the rectifier through the DC link capacitor. The load voltage at the point of common coupling U_L can be maintained without interruption. So the load can operate normally with little influence by the supply voltage sag. In the process of voltage compensation, the proposed control method could limit the fluctuation of DC link voltage to less than 6%.

As show in Fig. 4.9b, three-phase unbalance occurs between 0.5 s and 0.6 s. The voltage of phase-A is not changed; the voltage of phase-B drops 1 kV; the voltage of phase-C rise 1 kV; and phase angle keeps invariant. The DF-VQC put into operation at 0.5 s. As the DF-VQC can be operated at phase-splitting compensation mode, the three-phase inverter of DF-VQC can output unbalance voltage to compensate the supply voltage. When the DF-VQC is adopted, U_L almost keep unchanged, and the DC link voltage has a fluctuation when power grid voltage swells or sags, but soon will be able to stabilize around 800 V.

4.4.2.2 Fault Current-Limiting Function of DF-VQC

In order to show the fault current-limiting process and recovery process, simulations of DF-VQC for transient single phase to ground fault and transient short-circuit faults are carried out. Steady-state simulation results for single phase to ground fault (A), phase-to-phase short-circuit fault (A, B), two phase to ground fault (A, B), and three phase to ground fault are shown in Fig. 4.10.

In Fig. 4.10a, single phase (Phase-A) to ground fault occurs at 0.5 s, and disappears at 0.6 s. As only the ungrounded neutral grid is discussed in this book, the fault current is determined by the neutral voltage and the capacitor of the line to ground. For a 10 kV grid, the capacitor can be neglected, so the fault current is nearly zero. And the load current is unchanged. According to the proposed control strategy, the DF-VQC operates in voltage compensation mode. As shown in Fig. 4.10a, when single phase to ground fault occurs at 0.5 s, the PCC voltage of the healthy phase rise to the line voltage, and the fault voltage is zero. The line voltage of PCC and the load current keep unchanged. As the shunt transformers are delta-connected, the voltage on the low-voltage side of the shunt transformers is not influenced. So the DC link voltage

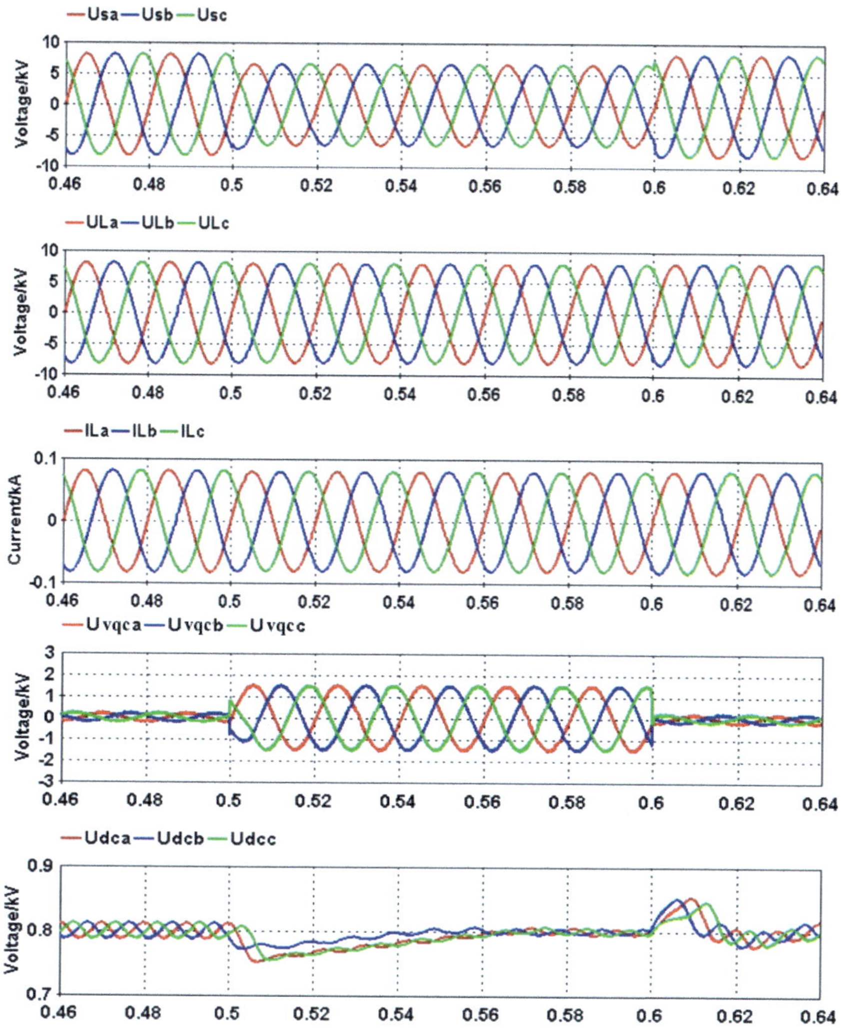

(a) Voltage fluctuation compensation

Fig. 4.9 Simulation results of voltage compensation

is stable when single phase to ground fault occurs. And the healthy phases are not interrupted. After the single phase to ground fault disappears at 0.6 s, the load current returns to be balance. DF-VQC still operates in the voltage compensation mode, and the DC link voltage is stable.

In Fig. 4.10b, phase-A to phase-B short-circuit fault occurs between 0.5 and 0.6 s. When the short-circuit fault was detected by DF-VQC, DF-VQC will switch to fault current-limiting mode. From Fig. 4.10b we can see that when short-circuit

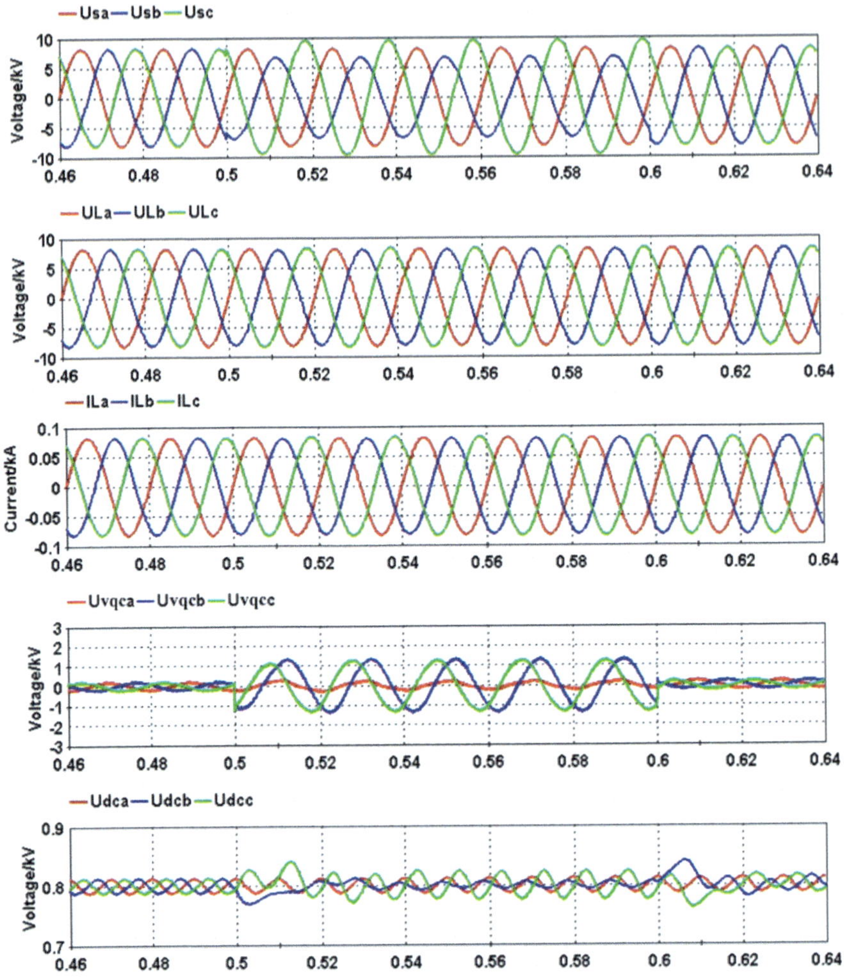

(b) Unbalanced voltage compensation

Fig. 4.9 (continued)

fault occurs, the load current of the fault phase will increase immediately. But when the DF-VQC is put into operation, the fault current of phase-A and phase-B can be limited to 880.4 A in about two cycles with an inrush current around 2 kA (Peak current), which is related to the time of putting into operation. And the steady-state fault current is determined by the phase-to-phase voltage, the equivalent impedance of the grid, the equivalent impedance of transmission line, and L. Figure 4.10b also illustrates that only the fault phases (Phase-A and Phase-B) are controlled by the DF-VQC, and the healthy phase (Phase-C) is not interrupted. The healthy phase

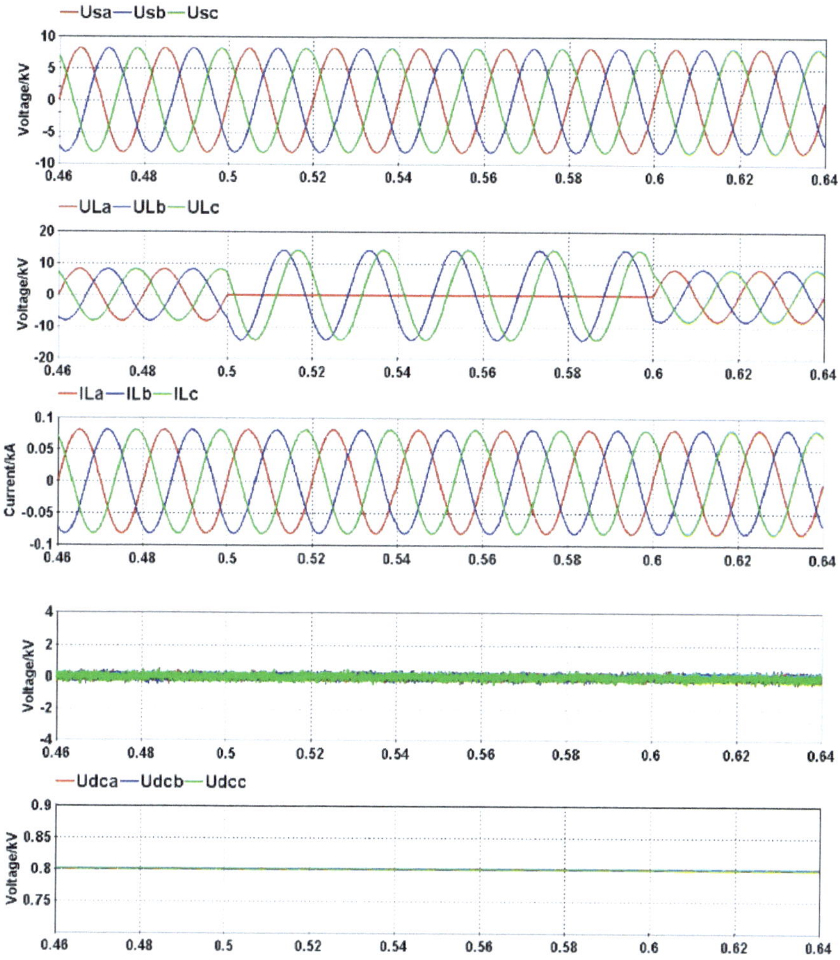

(a) Transient single-phase to ground fault current limiting.

Fig. 4.10 Steady-state simulation results of fault current limiting

could operate normally. When the DF-VQC is put into operation at 0.5 s, the DC link voltage rises less than 3%, which means the DF-VQC can word stably in the limiting process. When the fault disappears at 0.6 s, the fault phase of DF-VQC switches to voltage compensation mode immediately, with a 5% rise of DC link voltage. The dynamic limiting process and recovery process will be discussed in detail in the dynamic simulations.

In Fig. 4.10c, phase-A and phase-B to ground fault occurs between 0.5 and 0.6 s. Simulation results indicate that DF-VQC can immediately switch to fault current-limiting mode while the fault happens at 0.5 s, and switch to voltage compensation

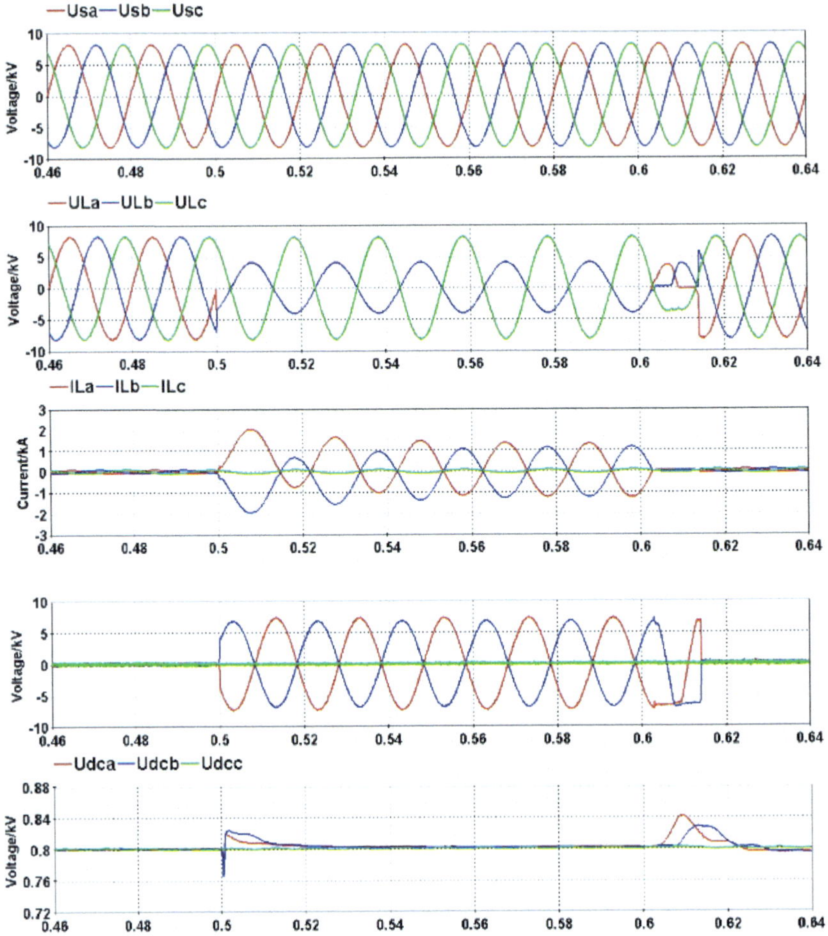

(b) Transient phase-to-phase short circuit fault limiting

Fig. 4.10 (continued)

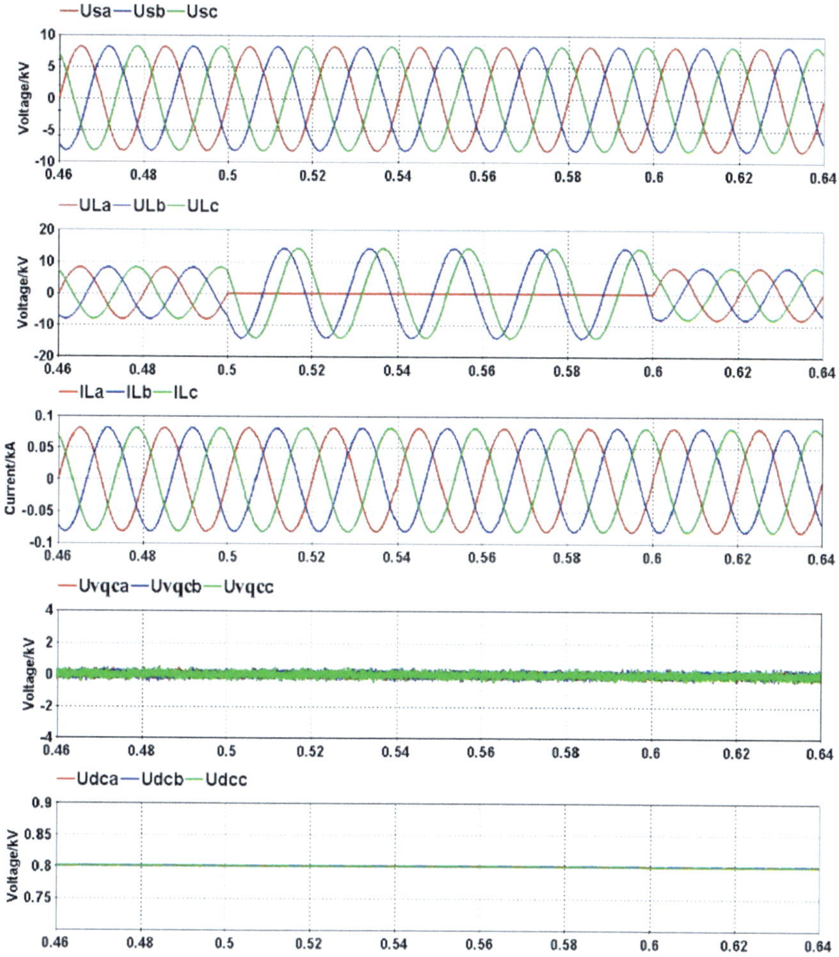

(c) Transient two-phase to ground short-circuit fault limiting

Fig. 4.10 (continued)

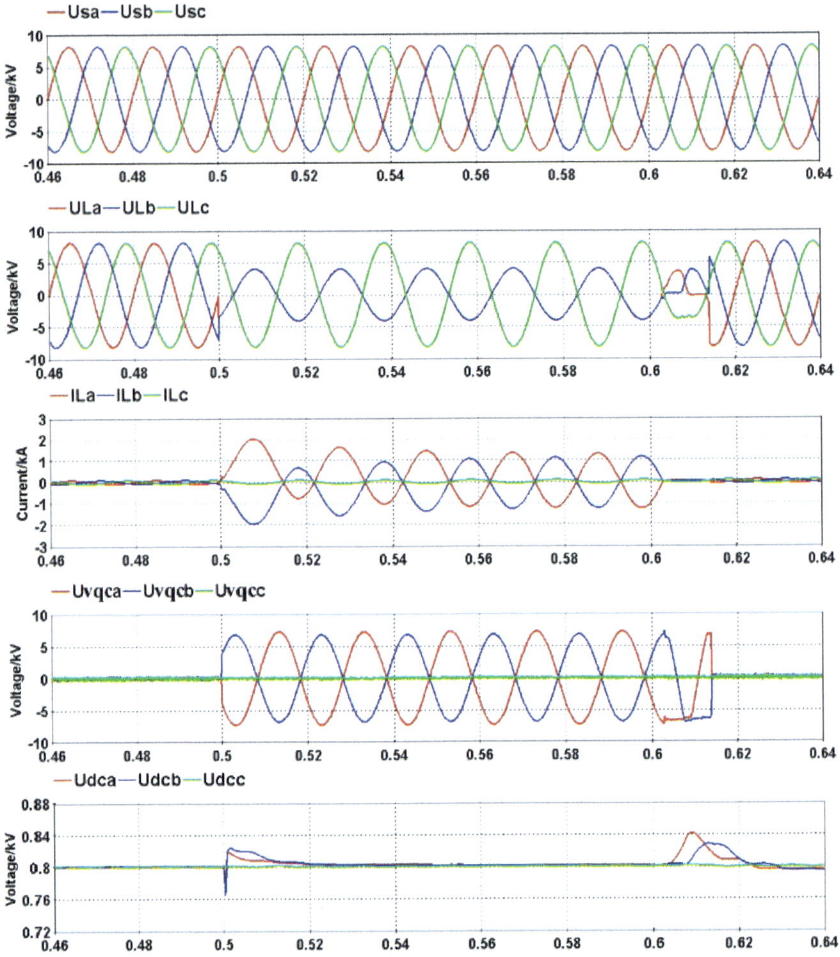

(d) Transient three-phase to ground short-circuit fault limiting

Fig. 4.10 (continued)

mode at 0.6 s when the fault disappears. The short-circuit current is limited to 903.1 A within two cycles. And the healthy phase (Phase-C) could operate normally with little influence by the fault current-limiting process. The DC link voltage rises less than 3 and 0.5% during the limiting process and recovery process.

In Fig. 4.10d, three phases to ground fault occurs between 0.5 and 0.6 s. Similarly as phase-to-phase short-circuit fault and two-phase to ground fault, the DF-VQC can immediately switch to fault current-limiting mode while the fault happens, and also can switch to voltage compensation mode while the fault disappears. The short-circuit current is limited to 1016.9 A within two cycles, and the DC link voltage is stable with less than 3% rise in the limiting and recovery process.

From Fig. 4.10, we can see that DF-VQC could realize a good steady performance when different kinds of transient short-circuit fault happen. To show the dynamic characteristics of the limiting process and recovery process, dynamic simulations are carried out. As shown in Fig. 4.10, the limiting and recovery process of different faults are similar, we only three phase to ground short is considered.

Dynamic simulation results of fault current limiting and recovery process are shown in Fig. 4.11. In the simulation, short-circuit fault occurs between 0.5 and 0.6 s. To simulate the dynamic process exactly, time delay of fault current detection Δt_1, time delay between the fault happens and activating of SCR Δt_2, and time delay of SCR deactivating Δt_3 are considered, where $\Delta t_1 = 0.508$ ms, $\Delta t_2 = 1$ ms, $\Delta t_3 =$ 14 ms.

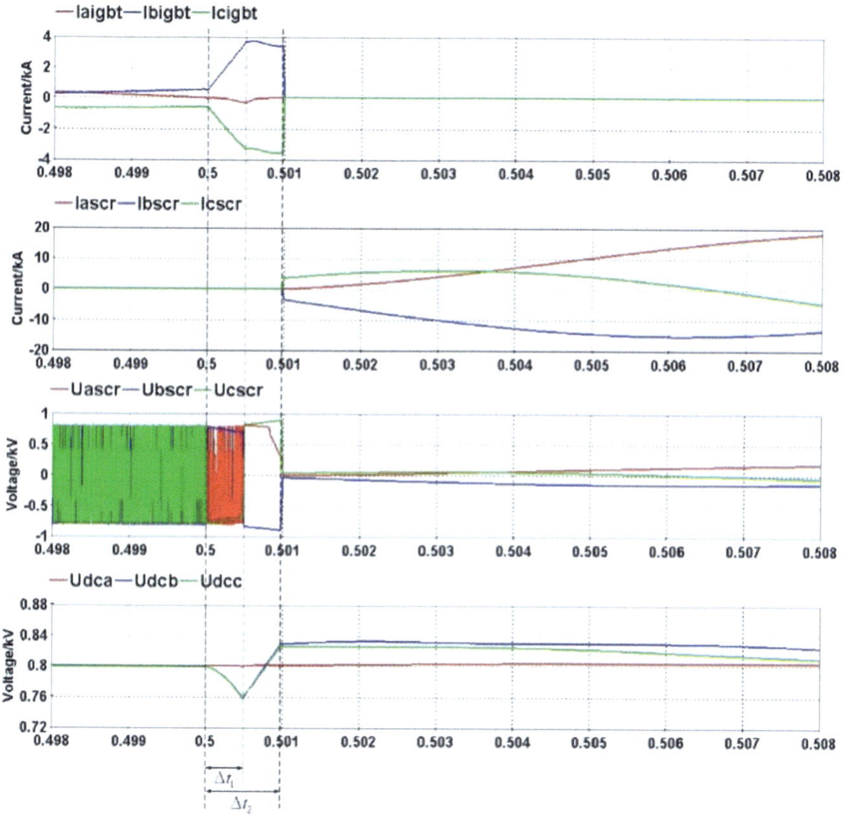

(a) Dynamic limiting process of three-phase to ground short-circuit fault

Fig. 4.11 Dynamic simulation results of fault current limiting

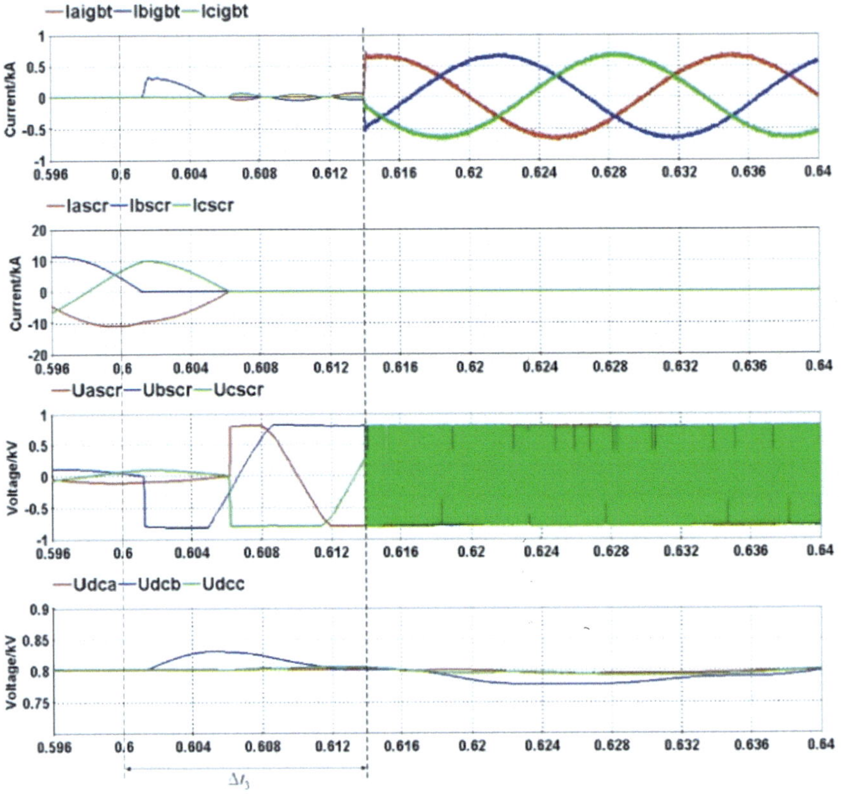

(b) Dynamic recovery process of three-phase to ground short-circuit fault

Fig. 4.11 (continued)

As shown in Fig. 4.11a, three phase to ground short-circuit fault occurs at 0.5 s, IGBT of DF-VQC is deactivated after Δt_1 (Δt_1 denotes the time delay of short-circuit current detection, in this book $\Delta t_1 = 0.508$ ms) delay. During Δt_1, the current of IGBT rise to nearly 4 kA (peak value). After Δt_1 delay, IGBT is deactivated and the current of IGBT decreases, but it doesn't become zero immediately. That is because during the period of 0.500508 and 0.501 s, the inverter of DF-VQC sustains the supply voltage. So the supply voltage will reverse charging the DC-side capacitor through the fly-wheel diode of IGBT, which will influence the stability of DC link voltage. Until the SCR is activated at 0.501 s, the output voltage of the inverter is zero, no current flows in the IGBTs. When SCR is activated, short current is limited by the L of the LC output filter. From the dynamic characteristic of the fault current-limiting process, the longer the time delay of fault current detection Δt_1 and activating of SCR Δt_2, the larger the inrush current of the IGBT, so fast fault current detection method is necessary.

As shown in Fig. 4.11b, when the short-circuit fault disappears at 0.6 s, the control signal to deactivate the SCR will be generated. To deactivate the SCR, a reverse voltage should be also sustained by the SCR. So the SCR is deactivated until 0.606 s. At 0.614 s, IGBT of DF-VQC is activated, and the DF-VQC operates in voltage compensation mode. So from the recovery process, it can be seen that the recovery process will last 10 ms at least due to the characteristic of SCR.

4.4.3 Scaled-Down Experimental of Fault Current-Limiting Function

To show the fault current-limiting performance of the DF-VQC, a scaled-down single-phase experimental platform was developed in the laboratory. In this experiment, the supply voltage is set at 110 V, the equivalent inductance of the grid at 5.4 mH, the capacitance and inductance of the output filter at 20 μF and 1.6 mH, respectively. The ratio of the series transformer is 2:1 with a 5 kVA capacity. The nominal load is a 20 Ω power resistor. A 5 Ω shorting resistor is connected in parallel with the load resistor to simulate a short-circuit fault. When the total current exceeds 25.3 A, we consider it to be a short-circuit fault. A TMS320F2812 digital processor is used as the controller. The PWM inverter uses IGBT module FF450R12ME4 with a switching frequency of 6.4 kHz as the switching devices while a MTC600A/12E thyristor is used as the crowbar bidirectional switch. Experimental results of DF-VQC for fault current limiting and recovery process are shown in Fig. 4.12.

It can be seen that the fault current is limited to 20.2 A by the DF-VQC in Fig. 4.12. From Fig. 4.12b, we can see that the limiting process takes about 10 ms, with little influence on the thyristor and the inverter. During the limiting process, the DC link voltage remains stable with little fluctuation.

Fig. 4.12 Experiment results of fault current-limiting action

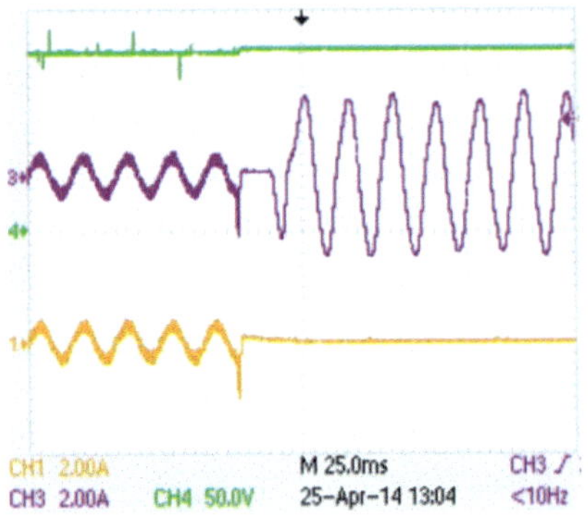

(a) Waveforms of supply current (CH3),DC linkvoltage (CH4) and the Output current of the inverter (CH1)

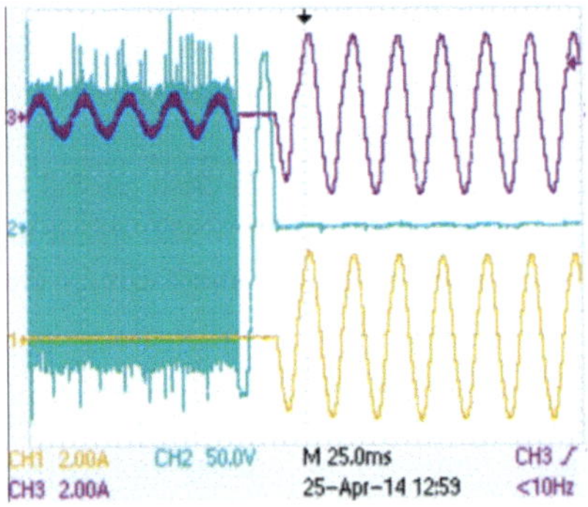

(b) Waveforms of current of the SCR (CH1),voltage sustained

by the SCR (CH2) and the supply current (CH3)

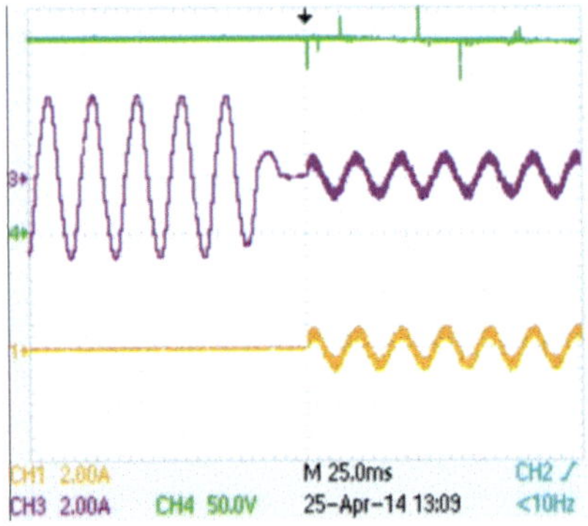

(c) Waveforms of supply current (CH3),DC link voltage (CH4), Output current of the inverter (CH1).

References

1. A.M.S. Atmadji, J.G.J. Sloot, Hybrid switching: a review of current literature, in *Proceeding of EMPD' 98*, vol. 2 (1998), pp. 683–688
2. A. Abramovitz, Irvine, K.M. Smedley, Survey of solid-state fault current limiters. IEEE Trans. Power Electron.**27**(6), 2770–2782 (2012)
3. A.R. Fereidouni, Amirkabir, I. Tehran, B. Vahidi, T.H Mehr, The impact of solid state fault current limiter on power netoperate with wind-turbine power generation. IEEE Trans. Smart Grid **4**(2), 1188–1196 (2013)
4. D.M. Divan, Inverter topologies and control techniques for sinusoidal output power supplies, in *Applied Power Electronics Conference and Exposition, APEC.91. Dallas* (1991), pp. 81–87

Chapter 5
Dual-Functional Voltage Quality Controller Based on Uncontrolled Rectifier Bridge Multiplexing

The second type of DF-VQC described in this section can operate in two main modes: short-circuit fault current-limiting mode and voltage fluctuation compensation pattern. It is mainly through the strategic control of the braking chopper circuit together with the anti-shunt thyristor to achieve a variety of functions, such as the ability to limit fault current.

5.1 Introduction

With the escalating demands on distribution networks and the increasing power requirements, fault current magnitudes can surpass the interrupting capacity of standard CBs. Replacing these circuit breakers to accommodate higher fault currents can be prohibitively expensive. Fuses represent an alternative solution for interrupting fault currents without the need for sophisticated sensors and actuators. However, their primary limitation is that they are disposable components, necessitating replacement after every fault event. In light of these limitations, the incorporation of a bypass circuit has emerged as a protective measure for series compensators during faults. This passive control strategy typically involves a fast-acting mechanical switch or a solid-state switch. While effective in diverting fault currents away from the VQC, this method alone cannot mitigate high fault currents, which might still pose a risk to other components within the power system, as highlighted in Ref. [1]. The challenge, therefore, lies in developing a comprehensive protection strategy that can limit fault currents, safeguard the VQC, and ensure the overall resilience of the power system against such events. This may involve advanced switching technologies, coordinated control systems, or integrated protection schemes that can respond swiftly and effectively to fault conditions.

F. Jiang et al., *Advanced Voltage Quality Controller for New Power Distribution Systems*, https://doi.org/10.1007/978-981-97-9861-2_5

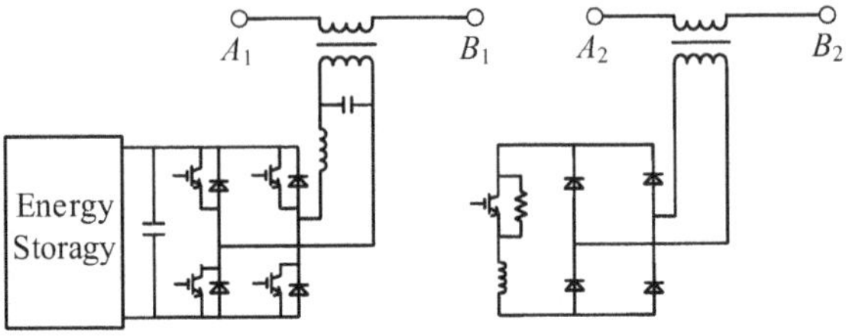

Fig. 5.1 Configurations of a typical. **a** Voltage quality controller and **b** bridge-type fault current limiter

A plethora of academic studies have recently explored various topologies for solid-state fault current limiters (SSFCLs), as noted in Ref. [2]. Among these, bridge-type solid-state fault current limiters (BSFCLs) have garnered attention for their compelling attributes, as highlighted in Ref. [3]. These advantages include minimal power dissipation during normal operation, rapid activation, swift recovery times, and the absence of a requisite fault detection subsystem, as detailed in Ref. [4]. Despite these merits, BSFCL designs often confront significant challenges. In the foundational BSFCL architecture, as discussed in Ref. [5], fault current confinement is achieved through the utilization of a superconducting coil. However, this configuration is limited in its capacity to sustain current limitation over prolonged periods, potentially necessitating the integration of a cumbersome and costly cooling apparatus for the direct current (DC) link.

An alternative strategy, illustrated in Fig. 5.1b and explicated in Ref. [6], involves the series connection of a resistor and an insulated gate bipolar transistor (IGBT) switch alongside the DC reactor. This setup allows for the modulation of fault current intensity by controlling the semiconductor switch's on and off intervals. Nevertheless, this method also entails the incorporation of a cooling system to safeguard the IGBT from thermal damage, as pointed out in Ref. [7].

Another proposition, as presented in Ref. [8], entailed the amalgamation of a BSFCL with a series current inverter, designed to absorb energy and restrain fault currents. While effective, this approach could result in elevated fabrication expenditures.

In a separate investigation, a new BSFCL topology was introduced, featuring six switching components as opposed to the customary four found in conventional BSFCL circuits, as per Ref. [9]. However, this augmentation in complexity may lead to a decline in switch efficiency.

Both VQCs and BSFCLs are recognized as pivotal components for enhancing the robustness and performance of power systems. However, the incorporation of such devices undeniably escalates the total investment required for power grid infrastructure. An arguably superior strategy would involve augmenting the inherent functionalities of VQCs and BSFCLs by endowing them with supplementary capabilities. Moreover, innovative control methodologies, as delineated in Refs. [10, 11], have been devised to restore voltages at the load side during upstream faults and simultaneously limit fault currents for downstream anomalies, utilizing a series compensator. Yet, the technique outlined in Ref. [12] presents a significant drawback, as it mandates the presence of a high-capacity energy storage system on the DC side, thereby introducing additional complexities and expenses. A noteworthy development has been the introduction of a fault current-limiting voltage quality controller (FCL-VQC), as proposed in Refs. [13, 14]. This device integrates the features of a VQC and an FCL, aiming to provide comprehensive protection and regulation. However, one critical shortcoming is identified in the design of the FCL-VQC; the output filter inductance is insufficiently sized to effectively curb fault current magnitudes, potentially undermining its current-limiting effectiveness. Concurrently, attention has been given to mitigating harmonic issues that may arise during the adjustment phase of fault currents.

Observing Fig. 5.1a, b, it becomes apparent that the VQCs exhibit striking similarities to BSFCLs. Both architectures feature diode bridge converters coupled with a series-connected injection transformer, wherein the bridges are interfaced with the secondary winding of the transformer. Nevertheless, the VQC's design is augmented with an LC filter, IGBTs, a DC-side capacitor, and an energy storage component—a configuration absent in the BSFCL, which is characterized solely by its current-limiting circuit. Drawing upon the synergistic attributes of these two designs, this book introduces a DF-VQC. This novel topology can be conceptualized as a fusion of the BSFCL and VQC, offering dual capabilities. The DF-VQC is proficient not only in executing voltage compensation akin to the VQC but also swiftly restrains short-circuit currents in the event of faults, emulating the BSFCL's functionality. Given the operational contexts of both the VQC and BSFCL, the DF-VQC demonstrates versatility, making it suitable for deployment across a spectrum of voltage levels, ranging from low (for instance, 400 V) to medium (such as 10 kV), as elaborated in Ref. [15]. Furthermore, the DF-VQC's potential applications extend beyond conventional settings, presenting promising opportunities in renewable energy sectors. Its utility is envisaged in scenarios such as photovoltaic power plants and wind farms, as detailed in Ref. [16], thereby underscoring its adaptability and potential impact across diverse energy landscapes.

The structure of this book unfolds as delineated below. In Sect. 5.2, we delve into the architectural blueprint of the DF-VQC, elucidating its dual-capacity design philosophy that amalgamates both voltage regulation and fault current limitation functionalities. Section 5.3 is dedicated to outlining a potent control strategy alongside a systematic methodology for selecting optimal parameters, ensuring the DF-VQC operates at peak efficiency and reliability. Section 5.4 undertakes a comprehensive examination of the operational chronology for the DF-VQC, detailing how it transitions through various states to fulfill its designated roles under differing conditions. Sections 5.5 and 5.6 are replete with simulation outcomes and empirical data garnered from tests conducted on the proposed topology, substantiating its theoretical efficacy and practical viability in real-world applications.

5.2 DF-VQC Topology

Illustrated in Fig. 5.2 is the innovative DF-VQC topology, merging the functionalities of the BSFCL and the VQC. This configuration is segmented into three primary components for each phase: an input section, a linking segment, and an output portion. At the input stage, a solid-state inverter (SM) interfaces with the electrical grid via a shunt transformer (for instance, T_a), accompanied by a filter (L_{za}) designed to mitigate high-frequency ripple effects. The output section mirrors a similar arrangement, where another inverter links to the grid through a series transformer (T_1), complemented by an LC filtering network. These two modules are bridged by a direct current (DC) link capacitor (C_{dc}), facilitating energy storage and voltage stabilization. A distinguishing feature of the DF-VQC, setting it apart from traditional VQCs, is the inclusion of a braking chopper circuitry and antiparallel thyristors (designated as SW1 and SW2). Within this braking chopper, a GTO is coupled in series with an IGBT, denoted as T, and a current-limiting inductor (L_1). Parallel to the IGBT, an adjustable resistor (R_1) is positioned to modulate the magnitude of the limiting current. Additionally, a discharge resistor (r_1) is connected in parallel with L_1 via a diode, serving to safely dissipate residual energy during operation. This arrangement not only enhances the versatility of the system but also ensures robust performance across a spectrum of operational scenarios, including fault mitigation and voltage conditioning tasks.

In Ref. [17], a bypass switch is configured in parallel alignment with the injection transformer. This switch, denoted as K, often comprises a pair of thyristors wired in parallel, selected for their capability to rapidly switch between conducting and nonconducting states. These thyristors are essential to have high enough current capacities to manage fault currents effectively, particularly when a fault triggers the need to bypass the transformer, as explained in Ref. [18].

The brake chopper circuit is strategically utilized to curtail the magnitude of fault currents. Meanwhile, the antiparallel thyristors serve to safeguard the DC capacitor

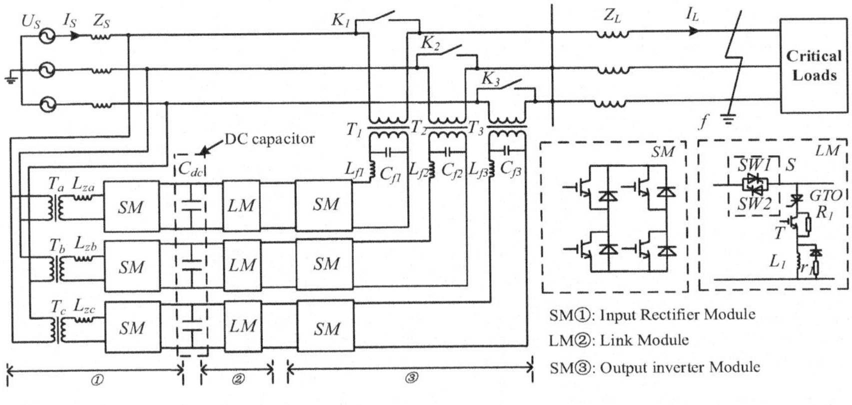

Fig. 5.2 Configuration of the proposed DF-VQC

and the energy storage device from being affected by the fault currents, thus ensuring the integrity of these components during critical situations. An integral part of the DF-VQC is the L_C filter, made up of inductor L_f and capacitor C_f, which works to mitigate various s undesired operational states that might otherwise interfere with the proper functioning of both the VQC and BSFCL modes. The design of the DF-VQC was carried out with careful consideration given to the specifications of the power electronic components, ensuring compatibility and reliability. The process of selecting the most suitable parameters for optimizing the performance of the DF-VQC is detailed in Sect. 5.3. This meticulous parameter selection is crucial for harmonizing the operational efficiency with the capabilities of the power electronic elements, ultimately contributing to the enhanced functionality and robustness of the DF-VQC.

5.2.1 Voltage Compensation Function of the DF-VQC

During regular operation, the switch K remains in an open state. Its principal purpose is to decouple the DF-VQC from the electrical system when the device is not in service or undergoing maintenance, as detailed in Ref. [19]. In circumstances where the source voltage exhibits imbalance or distortion, the DF-VQC transitions into its VQC mode of operation. This mode is specifically designed to counteract and correct any voltage irregularities present in the system. The simplified single-phase representation of the DF-VQC operating in voltage compensation mode is illustrated in Fig. 5.3.

The input module provides the required power to compensate for the voltage sags in the utility grid, as it injects voltage with suitable magnitude and phase in series with the line through series transformer T_1. The injection voltage is expressed as

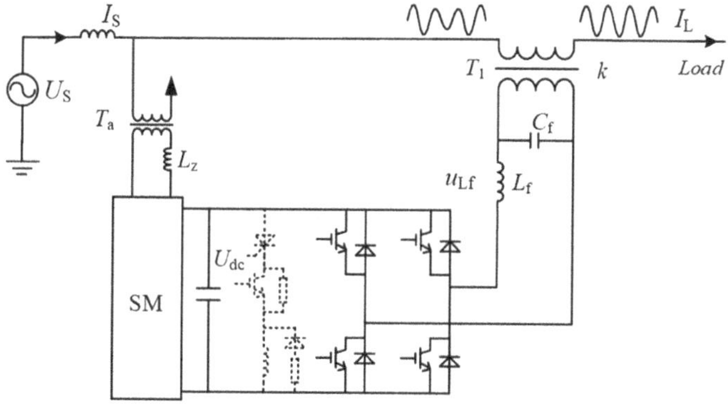

Fig. 5.3 DF-VQC system in the voltage compensation mode

$$u_{\mathrm{vqc}} = k(aU_{\mathrm{dc}} - u_{\mathrm{Lf}}), \tag{5.1}$$

where U_{dc} is the DC link voltage, α denotes the modulation depth, k represents the turns ratio of the series transformer, and u_{Lf} is the filter voltage. The inverter of each phase in the DF-VQC is operated and controlled independently, so that unbalanced voltage sags and swells may be compensated along with balanced ones. Moreover, some rating and design considerations of the DF-VQC injection transformer were taken into account to avoid saturation in Ref. [20]. These details are not discussed in this book for the sake of brevity.

5.2.2 Current-Limiting Function of the DF-VQC

Upon experiencing a short-circuit fault on the load side, there is an abrupt escalation in the line current. Upon detection of this fault, the IGBTs are promptly de-energized. The diode bridge converter, depicted in Fig. 5.4a, then acts as a current rectifier, converting the alternating current into direct current. Simultaneously, the trigger signals to the thyristors S are discontinued.

Fig. 5.4 DF-VQC in the current-limiting mode

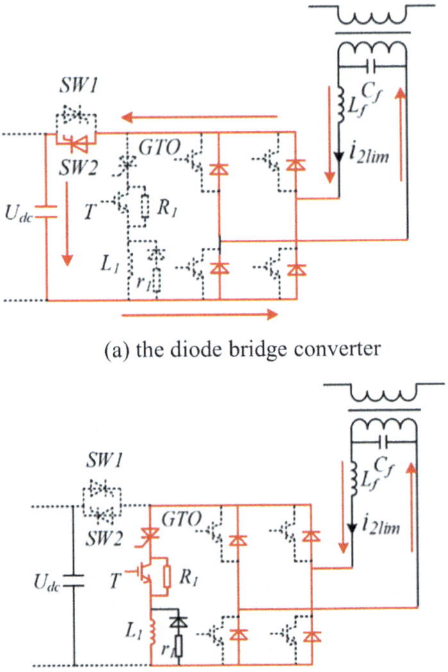

(a) the diode bridge converter

(b) the brake chopper circuit

At this juncture, the current exclusively flows through switch SW2. Consequently, SW1 is deactivated, whereas SW2 remains conductive until the current reaches its zero-crossing point. Following this, the GTO is activated, enabling the brake chopper circuit to redirect the fault current away from SW2 and assist in its shutdown. Components R_l, L_l, and L_f are introduced into the distribution line through transformer T_1.

By judiciously controlling the duration for which switch T is either on or off, the fault current can be restricted to a manageable level. This ensures that the system is protected from the deleterious effects of excessive currents during fault conditions. The single-phase equivalent circuit of the DF-VQC operating in current-limiting mode is illustrated in Fig. 5.4b, providing a clear visual representation of current pathways and protective mechanisms engaged during critical operational phase.

The limiting fault current is given by

$$I_{\mathrm{lim}}(t) = \frac{U_S(t)}{Z_S + Z_{\mathrm{lim}}} \approx \frac{U_S(t)}{Z_{\mathrm{lim}}}, \tag{5.2}$$

where Z_{lim} is the limiting impedance. Given that $Z_{\mathrm{lim}} \gg Z_S$, I_{fault} is mainly determined by the value of Z_{lim}. We may ignore the leakage reactance and the resistance of T_1, in which case, Z_{lim} is determined by R_l, L_l, L_f, C_f, and k. In addition, given that C_f is usually a small value, $|1/j\omega C_f|$ is typically a large value, so it may be concluded that the limiting impedance reads

$$|Z_{\mathrm{lim}}| \approx k^2\{|(1-\beta)R_l + j\omega L_l + j\omega L_f|//|1/j\omega C_f|\}, \tag{5.3}$$

where β is the duration of on/off periods of the switch T, and $|1/j\omega C_f| \gg |(1-\beta)R_l + j\omega L_l + j\omega L_f|$. Hence, I_{lim} is mainly determined by $|(1-\beta)R_l + j\omega L_l + j\omega L_f|$ and can be limited to an approximate value by adjusting the on–off duration of the switch T.

5.3 Control Scheme and Parameter Selection

The DF-VQC may operate in different modes, such as VQC and BSFCL modes, according to the state of the grid. In this section, we explain in detail the control scheme and optimal parameter selection.

5.3.1 Control Scheme Design

Up to now, several studies have been conducted on the control scheme of traditional VQCs in reference. In this book, we use the voltage compensation control strategy

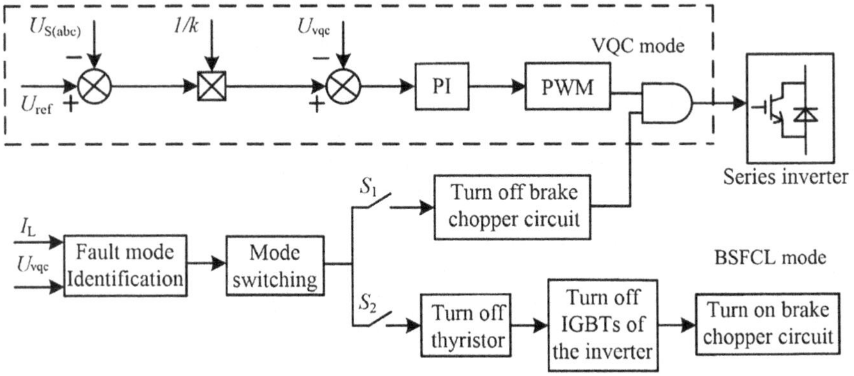

Fig. 5.5 Proposed DF-VQC control scheme

for a DF-VQC system, as described in Ref. [21]. This control scheme is shown in Fig. 5.5.

Figure 5.6 illustrates a graphical representation of a fault detection technique specifically designed for current monitoring. This methodology, outlined in Ref. [13], is capable of detecting the load current and its rate of change. Here, Δt signifies the detection window for fault current, i_L and i_{L_fault} represent the line currents under normal and fault conditions, respectively, while Δi_L and Δi_{L_fault} denote their corresponding rates of change in these scenarios. This approach notably excels in addressing two particular scenarios: Firstly, in instances where a fault arises precisely at the zero-crossing point of the line current, the fault current momentarily equals zero, yet its rate of change remains nonzero. Secondly, when the fault occurs at the peak value of the line current, the rate of change momentarily vanishes, yet the fault current itself remains nonzero.

Here, Δi_L and Δi_{L_fault} signify the differences in the immediately succeeding sampling points between the load current and the fault current, respectively. Upon detection of a short-circuit fault, the DF-VQC promptly transitions to the BSFCL

Fig. 5.6 Fault detection method of the DF-VQC

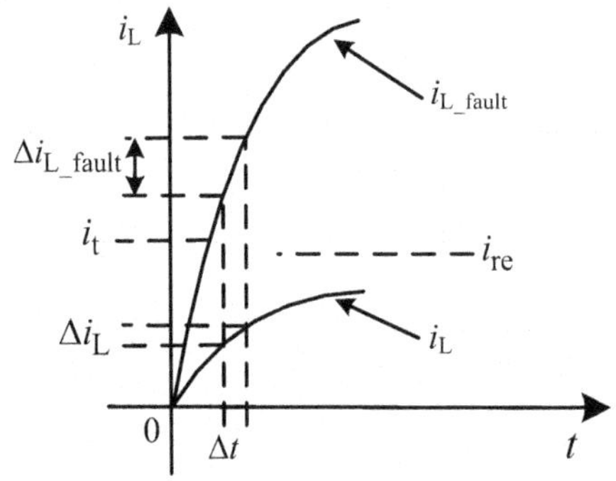

mode, enabling swift containment of the fault current to a safe level through modulation of the switching duty cycle β of the transistor T. This seamless mode transition takes place within milliseconds, ensuring that the current-limiting function engages prior to the initial peak of the line current. In this context, ire represents the minimum triggering current for the inverse time overcurrent protection device, set to exceed the peak load current. Specifically, ire is calculated as 1.2 times the square root of 2 multiplied by irate, where irate is the root mean square (RMS) value of the rated load current.

5.3.2 Optimal Parameter Selection

(1) Switch design elaboration: The braking chopper serves a crucial function in restricting the fault current to a predetermined level. Furthermore, the system incorporates four antiparallel diodes within the inverter's upper section, which are tasked with managing the restricted currents efficiently. In scenarios where an overcurrent condition arises due to a short-circuit fault, the pulse width modulation (PWM) control mechanism may be deactivated, subsequently switching off all IGBTs within the series converter. Consequently, the fault current redirects its flow through the converter's diodes. When the switch T is conducting, this may be described by

$$u_{s2} = u_{vqc}/k = i_{2\,\mathrm{lim}}(Z_{\mathrm{lim}}/k^2), \tag{5.4}$$

where u_{VQC} and u_{S2} represent the primary and secondary voltages of the series transformer during the faults, respectively, and $i_{2\mathrm{lim}}$ is the secondary current of the series transformer during the faults. As can be seen from (5.3) and (5.4), the upper limit of $i_{2\mathrm{lim}}$ is related to the maximum permissible current of the GTO, the switch T, and the four antiparallel diodes. Thus, we have

$$i_{2\,\mathrm{lim}} \leq \min(i_{\mathrm{T}}, i_{\mathrm{GTO}}, i_{\mathrm{D}}), \tag{5.5}$$

where i_{T}, i_{GTO}, and i_{D} denote the maximum permissible currents of the T, GTO, and diodes, respectively. Moreover, the rated blocking voltage of each switching component should satisfy the condition

$$u_{S2} \leq \min(u_{\mathrm{T}}, u_{\mathrm{GTO}}, u_{\mathrm{D}}), \tag{5.6}$$

where u_{T}, u_{GTO}, and u_{D} represent the corresponding rated blocking voltages of the switch, GTO, and diodes.

(2) The series transformer, a staple in traditional designs of VQCs and BSFCLs, plays a dual role. Not only does it mitigate the voltage demand on the inverter, but

it also ensures galvanic isolation between the inverter and the power grid. Practical considerations, however, necessitate attention to the thermal equilibrium and maximum voltage endurance of the current-limiting component, making the determination of the ideal transformer turns ratio a multifaceted challenge. For the sake of conciseness, we refer to the methodology outlined in Ref. [14] for this selection process, which has been adopted in our study. To avert transformer saturation, the design principles and operational guidelines detailed in Ref. [21] were implemented. When the DF-VQC is engaged in its BSFCL function, the fault current traverses the LC filter and the primary transformer T_1. Alternatively, in the realm of series transformer design for VQCs and BSFCLs, the transformer not only eases the voltage stress on the inverter but also establishes a protective barrier against direct electrical coupling with the grid. The practicalities of ensuring both thermal robustness and high voltage tolerance in the current-limiting element add layers of complexity to selecting the most suitable series transformer ratio. We have adhered to the approach detailed in Ref. [14], which we do not reiterate here to maintain brevity. To safeguard against transformer saturation, the design and operational strategies described in Ref. [21] were utilized. When the DF-VQC assumes its BSFCL role, the LC filter and transformer T_1 become the path through which the fault current travels. Then, T_1 withstands the supply voltage during the faults, and while its capacity should satisfy the condition given by

$$S_{\mathrm{T}} \geq U_S I_{\mathrm{L\_fault}}. \tag{5.7}$$

In addition, the thermal stability of L_{f} and maximal tolerance voltage of C_{f} should be considered during the faults in Ref. [13]. This may be expressed as

$$\begin{aligned} Q_{Lf} &\geq \int_{b}^{a} \left(k i_{\mathrm{L\_fault}}\right)^2 dt \\ a &= t_{\mathrm{return}} \\ b &= t_{\mathrm{fault}}, \end{aligned} \tag{5.8}$$

where Q_{Lf} indicates the thermal stability of L_{f} during the faults, and t_{fault} and t_{return}, respectively, denote the times at which the fault occurs and is restored. Given that C_{f} will be easily damaged by the overvoltage, U_{Cf} may be expressed as

$$U_{\mathrm{Cf}} \geq \max(u_{\mathrm{S\,max}}/k, u_{\mathrm{dc}}), \tag{5.9}$$

where U_{Cf} is the maximal voltage rating of the filter capacitor, u_{dc} denotes the DC voltage, and u_{Smax} represents the primary voltage of transformer T_1 in Ref. [14].

(3) Current limitation capability: The upper bound of the fault current's amplitude is governed by the key elements of the brake chopper circuit, namely L_{f} and k. This allows for the short-circuit current to be effectively constrained to a

desirable level through the strategic use of L_1, R_1, L_f, k, and β. Notably, L_f and k play a crucial role in the compensation mechanism of the DF-VQC, with k being derived from Eq. (5.1). Furthermore, the determination of L_f and C_f can be facilitated by referring to the methodology detailed in Ref. [14].

$$\left\{\frac{L_f = \rho/2\pi f_c}{C_f = L_f/\rho^2 = 1/2\pi f_c \rho}\right., \tag{5.10}$$

where ρ and f_c denote the damping factor and resonance frequency, respectively. First, the compensation ability of the DF-VQC should be determined using L_f and k. When these values have been selected, the amplitude limit of the fault current is mainly adjusted using L_1, R_1, and β. When T is conducting during a fault, the maximum value of the limited current, $i_{\text{lim\_max}}$, may be calculated as

$$i_{\text{lim\_max}}(\beta = 1) = \frac{U_S(t)}{Z_S + k^2(\omega L_1 + \omega L_f)}. \tag{5.11}$$

When the IGBT is completely turned off during the faults, the minimum value of the limited current, $i_{\text{lim\_min}}$, is calculated using the equation

$$i_{\text{lim\_min}} \leq i_{\text{lim}}(\beta) \leq i_{\text{lim\_max}}, i_{\text{lim\_max}}(\beta = 0) = \frac{U_S(t)}{Z_S + k^2(R_1 + \omega L_1 + \omega L_f)}. \tag{5.12}$$

Thus, the limited current should meet the following condition:

$$i_{\text{lim\_min}} \leq i_{\text{lim}}(\beta) \leq i_{\text{lim\_max}}. \tag{5.13}$$

(4) In the context of safeguarding the DC link capacitor within power electronics systems, two principal strategies are outlined in the literature to ensure its integrity under various fault conditions. Firstly, in scenarios where a short-circuit fault occurs on the load side, an immediate response mechanism is activated through the control of a dedicated switch, designated as S. By promptly deactivating this switch, the DC link capacitor is electrically decoupled from the faulty section, thereby mitigating the risk of damage due to fault currents. This preemptive action is critical in preserving the capacitor's operational lifespan and system reliability.

Conversely, in situations devoid of severe faults like short circuits, yet susceptible to less dramatic but still impactful voltage fluctuations or load variations, such as those induced by motor startups or capacitor switching, the role of varistors becomes pivotal. These nonlinear resistors act as voltage surge protectors, effectively clamping any transient over voltages that could otherwise stress the DC link capacitor. Notably, during the initial stages of a short-circuit fault before the protective switch can react, varistors provide an immediate barrier against excessive voltages, bridging the gap between fault detection and switch disengagement. The threshold value of the DC link protection voltage, U_{dc_th}, is set to $1.25U_{dc}$, where $U_{dc} = 200$ V and $U_{dc\_th} = 250$ V. Thus, the DF-VQC presents fast switching into the fault current-limiting function.

5.4 Operation Sequence Analysis

During routine operation, the DF-VQC functions akin to a standard VQC. Under normal circumstances, devoid of any disturbances, the switch K remains open, allowing the system to operate seamlessly. However, upon sensing voltage sags or the incursion of short-circuit fault currents, the DF-VQC swiftly shifts its paradigm, transitioning into either the fault current-limiting mode or the voltage compensation mode, thereby enhancing system resilience.

The book elaborates on the fault current-limiting protocol adopted by the DF-VQC in instances where a malfunction necessitates the switch from VQC mode to the BSFCL mode. This protocol ensures a seamless and effective response to mitigate the impact of faults. Furthermore, acknowledging the residual voltage instabilities post-fault, the document delineates a recovery procedure. This involves a strategic shift from the BSFCL mode back to the VQC mode, facilitating a smooth restoration of normal operational parameters during the recovery phase.

Figure 5.7 encapsulates the operational logic of the DF-VQC, illustrating how it navigates through different modes based on real-time system conditions, underscoring its adaptability and efficiency in managing both steady-state operations and transient disturbances.

5.4.1 *Fault Current-Limiting Scheme Sequence*

The operational flowchart for the fault current limitation scheme is presented in Fig. 5.8. During standard operating conditions, the line current (i_L) maintains a stable level within its normal range, defined by its effective value. However, a short-circuit event at T_1 causes an abrupt and significant increase in the line current, potentially endangering the power elements of the DF-VQC.

At the instant T_2, when the line current's magnitude surpasses a predetermined safety threshold, which is governed by the line relay protection settings as shown in Fig. 5.6, the system detects a fault current. Given the practical limitations of switches, there is an inherent delay in transitioning between the VQC and BSFCL operational modes. To mitigate this, rapid detection methods and efficient control mechanisms are paramount for the DF-VQC.

The process commences with the deactivation of the IGBTs in the series inverter at T_3. Subsequently, at T_4, the trigger signals to the antiparallel thyristors are discontinued. A deliberate time gap (ΔT_1, encompassing 2–3 sampling intervals) is maintained between T_3 and T_4 to ensure that the IGBTs are fully de-energized.

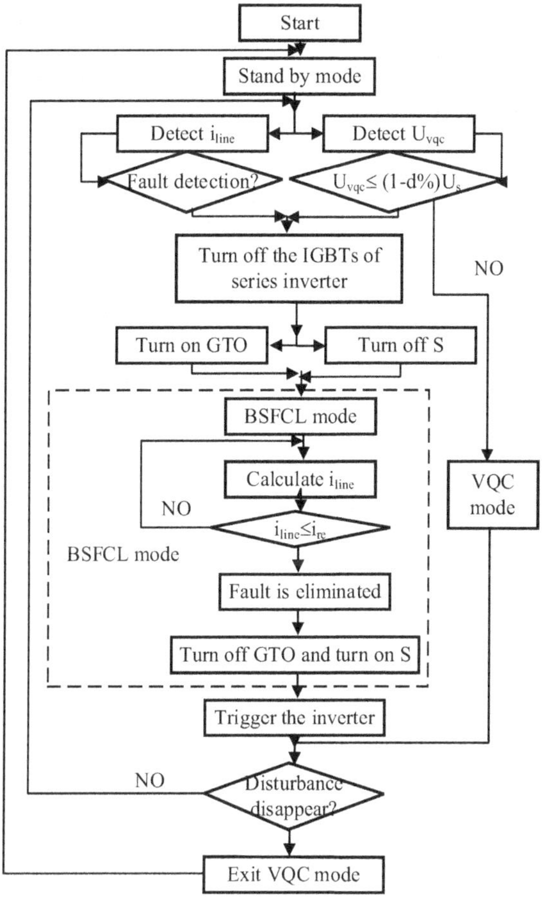

Fig. 5.7 Operational algorithm of the DF-VQC for different modes

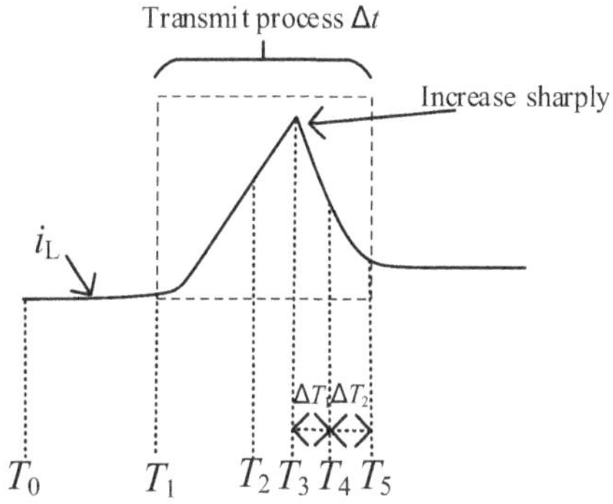

Fig. 5.8 Fault current-limiting scheme sequence

At T_5, the DF-VQC's control system initiates a trigger signal to the GTO thyristors, which possess swift response times for both turning on (10 μs) and turning off (25 μs), facilitating the prompt activation of the brake chopper circuit, as referenced in [22]. This action signifies the DF-VQC's transition into the BSFCL mode.

A further time delay, ΔT_2, exists between T_4 and T_5 to confirm the successful deactivation of SW1. Thus, the total duration for completing the transition to the BSFCL mode is represented by ΔT, calculated as the difference between T_5 and T_1. Despite the inherent delays in the transition process, they do not compromise the DF-VQC's ability to execute its intended functions effectively.

5.4.2 Recovery Scheme Sequence

Once the control system senses that the line current has diminished below the established threshold value (i_{re}) of the relay protection mechanism, signifying the elimination of the fault condition, a series of actions are promptly executed. The GTO thyristor is deactivated by applying an off-gate signal, ceasing its current-carrying function.

Simultaneously, the thyristor is commanded to turn on, establishing a conductive path, and IGBTs of the series inverter are re-triggered, restoring their operation. This coordinated sequence of events triggers the DF-VQC to shift from its BSFCL mode to its VQC mode.

In the VQC mode, the DF-VQC actively mitigates any voltage fluctuations that may have arisen during the fault recovery phase, ensuring the stability and reliability of the power supply.

5.5 Simulation Results

In this segment, we showcase simulation outcomes aimed at substantiating the practicality and effectiveness of the DF-VQC. The simulation, executed using the sophisticated PSCAD/EMTDC software suite, involved the circuit model depicted in Fig. 5.2. Here, the DF-VQC was strategically positioned in series, interconnecting the power source and the critical load, thereby playing a pivotal role in the system. The simulation was conducted under specific conditions, with detailed parameter values provided in Appendix A for reference. Notably, a source voltage of 110 V was applied, alongside a sensitive load characterized by an impedance of $20 + j5$ Ω. This setup allowed us to comprehensively evaluate the DF-VQC's performance under realistic operating scenarios, ensuring its capability to maintain voltage quality and limit fault currents as intended.

5.5.1 Voltage Compensation Function of the DF-VQC

The graphical illustrations of the DF-VQC's performance during voltage sags are presented in Fig. 5.9. Specifically, Fig. 5.9a demonstrates the swift transition of the DF-VQC into its voltage compensation mode in response to a voltage fluctuation occurring between 1.100 and 1.200 s, with a depth of 20%.

In Fig. 5.9b, it is evident that the load voltage (U_L) is effectively regulated to maintain a constant amplitude throughout this period, showcasing the DF-VQC's ability to stabilize the system under voltage sags.

Further, Fig. 5.9c reveals that the DF-VQC proficiently supplies the necessary compensation voltage, ensuring that the voltage fluctuations do not significantly impact the power system.

Moreover, Fig. 5.9d demonstrates that the load current (I_L) also remains stable at a constant amplitude, indicating that the DF-VQC's intervention maintains the system's operational integrity.

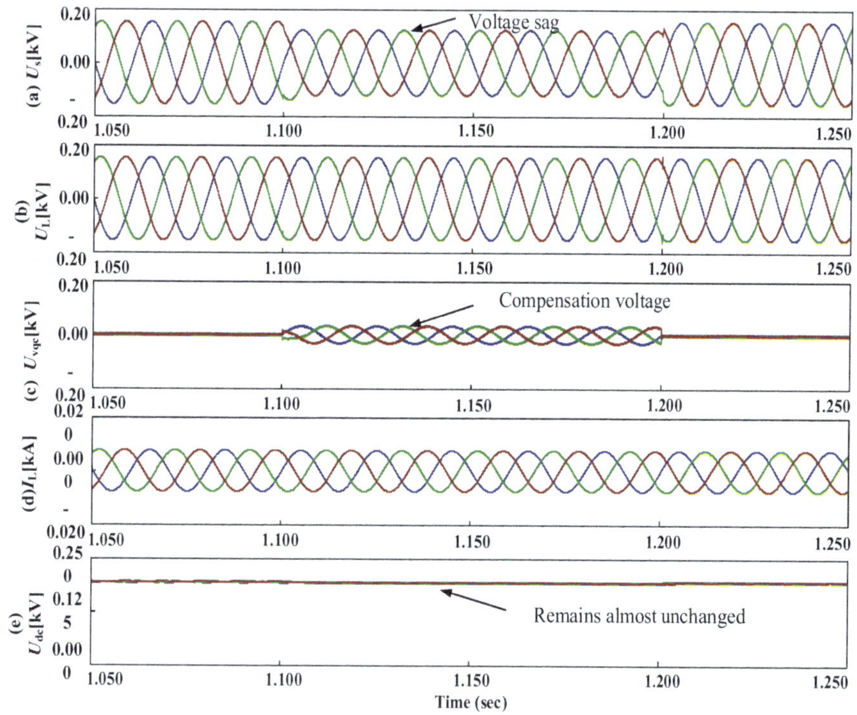

Fig. 5.9 Simulation results of the voltage compensation function of the DF-VQC system. Top to bottom: **a** Supply voltage U_S, **b** Load voltage U_L, **c** Primary voltage U_{VQC}, **d** Load current I_L, and **e** DC link voltage

Lastly, Fig. 5.9e highlights that the DC link voltage experiences minimal variations during the voltage sags, emphasizing the DF-VQC's effectiveness in preserving the stability of the power system's key parameters.

5.5.2 *Fault Current-Limiting Function of the DF-VQC*

During a short-circuit fault between phases A and B, occurring from 1.300 to 1.400 s, the line current experiences a rapid surge, exceeding a predefined threshold. This triggers the DF-VQC's faulty phases to switch into its BSFCL mode, taking into account the inherent delays associated with fault detection and mode transition. This includes the time required to shut down the input bridge, deactivate switch S, and activate the GTO thyristor, resulting in a transient current spike in the affected phases (A and B). Remarkably, the unaffected phase (C) continues to operate seamlessly during this event. As depicted in Fig. 5.10b, the load-side voltages of the faulted phases plummet close to zero during the fault, while the voltage in the healthy phase remains stable. Figure 5.10c reveals that the majority of the voltage drop across the faulted phases occurs across the series transformer of the DF-VQC, effectively limiting the fault current. When the switch operates at a 50% duty cycle, the limiting impedance (Z_{lim}) can be precisely calculated using Eq. (5.4). Figure 5.10d showcases the DF-VQC's effectiveness in swiftly constraining the load currents of the faulty phases to predetermined safe levels. Lastly, Fig. 5.10e emphasizes that the DC link voltage experiences minimal disturbance throughout the fault, demonstrating the system's robustness and stability.

Furthermore, we delved into the intricate transition dynamics triggered by the short-circuit fault. At the onset of the fault at time T_1, the currents in the faulty phases surge rapidly, accompanied by a slight decrease in the DC-side voltages, as evident in Fig. 5.11a, b. Subsequently, the IGBTs of the series inverter deactivate at T_3, whereupon the diodes within the inverter commence operation as a current rectifier. Concurrently, SW1 is promptly switched off, leading to a marginal increase in the DC-side voltage, as depicted in Fig. 5.11b.

Proceeding further, the trigger pulses to the antiparallel thyristors are withdrawn at T_4, paving the way for the GTO thyristor to commence conduction at T_5. This critical moment sees the GTO diverting the fault current, expediting the deactivation of SW2, as indicated by the noticeable inflection point in Fig. 5.11c. Following this sequence, the DC-side voltage regains its steady-state equilibrium, as the DC system becomes isolated from the power grid.

The timing diagram outlined in Fig. 5.11d illustrates the orchestrated sequence of events within the DF-VQC. Notably, despite the inherent delays encountered during the switching process, they do not compromise the DF-VQC's ability to execute its intended function effectively.

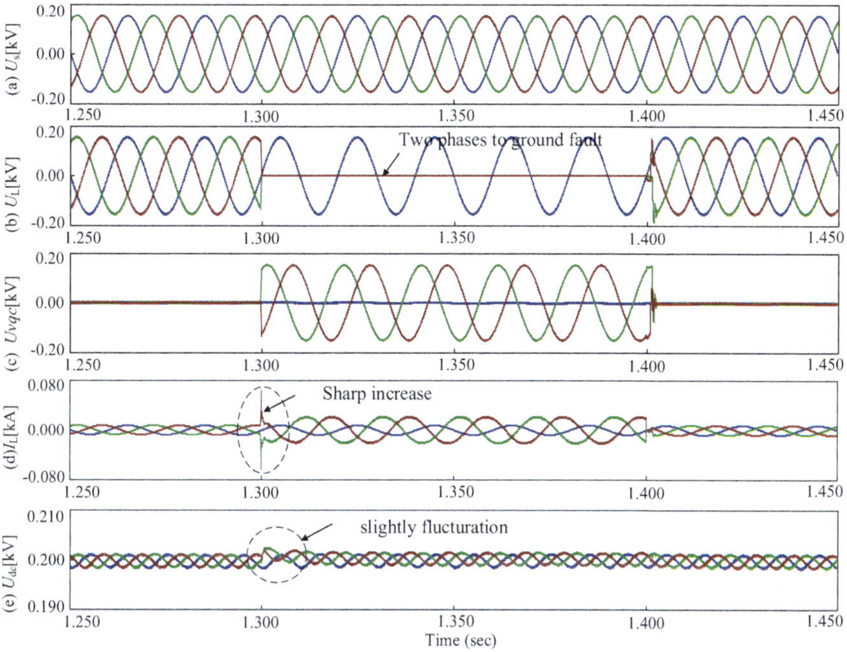

Fig. 5.10 Simulation results of the fault current-limiting function of the DF-VQC system during a two-phase downstream fault. Top to bottom: **a** Supply voltage U_S, **b** Load voltssage U_L, **c** Primary voltage U_{VQC}, **d** Load current I_L, and **e** DC link voltage

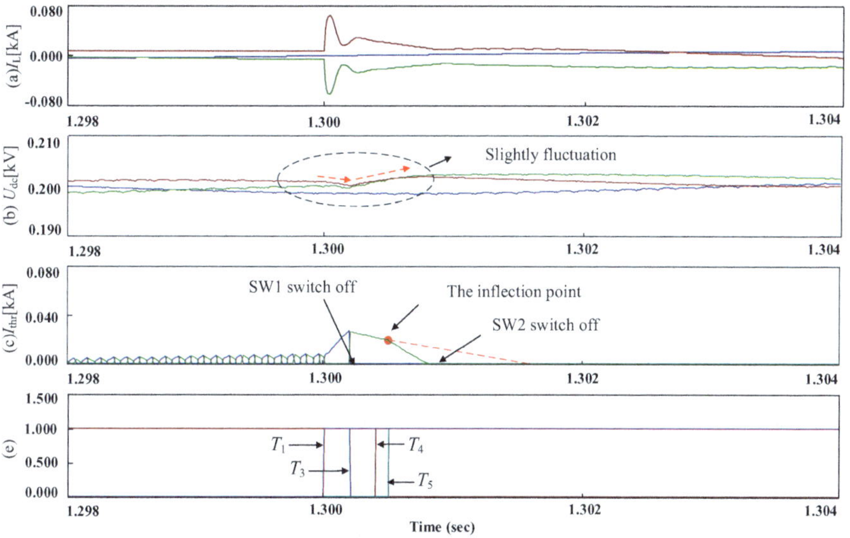

Fig. 5.11 Simulation results of the transient process of the fault current-limiting mode for the DF-VQC system. Top to bottom: **a** Load voltage U_L, **b** DC link voltage U_{VQC}, **c** the antiparallel thyristors I_{thr}, and **d** the timing sequence

5.5.3 Harmonics Current Analysis

Viewed from the perspective of distorted waveform mitigation, the DF-VQC enhances power quality through the deployment of a judicious operational strategy, particularly when functioning in VQC mode. The intensity of the fault current can be modulated by adjusting the duty cycle of the semiconductor switch, offering a degree of control over fault management.

However, when the DF-VQC transitions to its current-limiting mode, as noted in Ref. [23], the introduction of higher-order harmonic currents becomes inevitable, emanating from components such as the diode bridge converter, GTO, and IGBT. The analysis of harmonic content assumes paramount importance for the effective design of current protection mechanisms.

As depicted in Fig. 5.12, the harmonic contents of line currents during fault conditions have been quantified for various duty cycle settings of the switch, ranging from 20 to 80%. The corresponding harmonic percentages were found to be 1.41%, 1.53%, 1.72%, and 2.14%, respectively. It is noteworthy that as the duty cycle decreases, not only does the fault current's amplitude limit diminish, but the harmonic contents are also attenuated.

Despite the introduction of harmonics, the total harmonic distortion (THD) of the line currents remains within acceptable bounds, aligning closely with the limits stipulated by the IEEE519 standards. This suggests that the DF-VQC, when operating in BSFCL mode, exerts a negligible effect on line current harmonics, thereby ensuring compliance with industry standards for power quality.

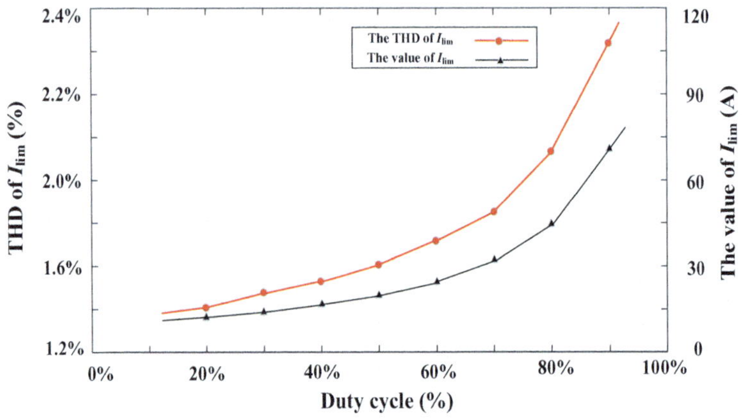

Fig. 5.12 Line current harmonics in the fault current-limiting conditions

5.6 Experiment and Analysis

As illustrated in Fig. 5.2, given that the DF-VQC system adopts a symmetrical three-phase architecture, and considering the constraints imposed by the interface compatibility between the real-time digital simulator (RTDS) and the controller, a single-phase experimental setup was employed. This setup served to effectively validate the operational capabilities and performance attributes of the DF-VQC's topological design. It is worth noting that all system parameters utilized in this experiment mirrored those detailed in Appendix A, ensuring consistency and comparability across different operational contexts.

5.6.1 Voltage Compensation Function of the DF-VQC

The performance of the DF-VQC in response to various voltage sags is showcased in Fig. 5.13. Prior to the events, the system operates in a stable, steady-state condition. Upon the occurrence of a voltage sag spanning from t_1 to t_2, characterized by a depth of 20% and a voltage undershoot (u_S) of 87.69 V over six cycles, the DF-VQC promptly transitions into its VQC mode, as depicted in Fig. 5.13a. This swift action enables the injection of a series voltage (u_{VQC}), effectively maintaining the desired load voltage (u_{Load}) at the load side. Consequently, as evident in Fig. 5.13b, the load current (i_{Load}) remains stable and unaffected throughout the duration of the voltage sags, while the DC link voltage (u_{dc}) experiences minimal deviation, essentially remaining unchanged.

5.6.2 Fault Current-Limiting Function of the DF-VQC

In the event of a fault, the load current instantaneously drops to zero, showcasing the DF-VQC's formidable fault current-limiting capability as illustrated in Fig. 5.14.

Case 1 ($\beta = 60\%$): At the moment of fault occurrence, denoted as time t_3, the load voltage (u_{Load}) diminishes to zero, whereas the load current (i_{Load}) spikes upwards, as demonstrated in Fig. 5.14 a. Notably, nearly the entire source voltage (u_{VQC}) is dissipated across the DF-VQC system, as evident from Fig. 5.14b. Throughout the fault duration, the current traversing the thyristors (i_{Thr}) dwindles to zero, whereas the current through the GTO thyristor remains nonzero. The isolation of the DC side, achieved through precise thyristor control, is evident in Fig. 5.14, where the DC voltage (u_{dc}) remains virtually unaffected and stable throughout the fault period.

Case 2 ($\beta = 20\%$ Scenario): At the instant t_5, a fault occurs, causing the load voltage (u_{Load}) to abruptly drop to zero. In response, the load current (i_{Load}) experiences a surge, yet remarkably, the DC voltage (u_{dc}) remains stable and unaffected throughout the fault event, as clearly shown in Fig. 5.14a. Meanwhile, Fig. 5.14b

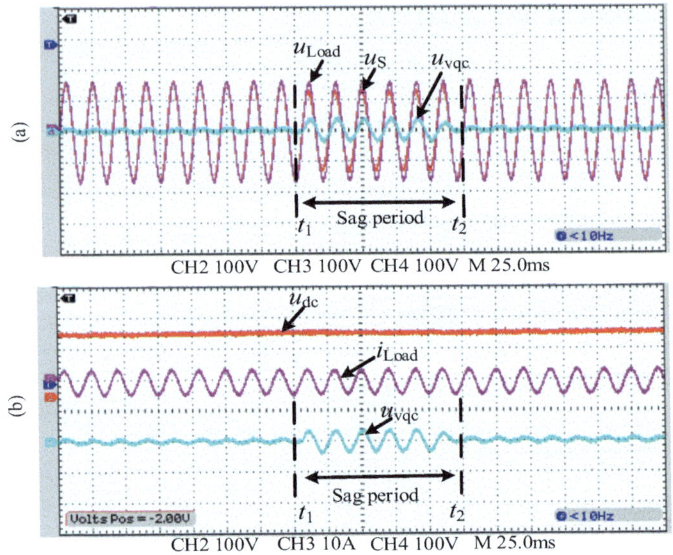

Fig. 5.13 Experimental results of the voltage compensation function of the DF-VQC system

illustrates that the current flowing through the thyristors (i_{Thr}) diminishes to zero, while the GTO continues to conduct a nonzero current. Notably, the control mechanism effectively manages the fault current (i_{Load}) by adjusting the duty cycle (β) of the switching element, with a gradual decrease in the peak value of the fault current as β decreases, evident in Fig. 5.14a, c.

The DF-VQC's exceptional fault current-limiting performance in scenarios involving near-maximum fault currents is further demonstrated in Fig. 5.15. Similar to the trends observed in Fig. 5.14, the behavior patterns in Fig. 5.15a–d are consistent, indicating the system's robustness. However, in comparison to Figs. 5.14a and 5.15a reveals a more pronounced initial surge in fault current at the onset of the fault, highlighting the DF-VQC's ability to swiftly and effectively dampen even severe fault conditions to maintain system stability.

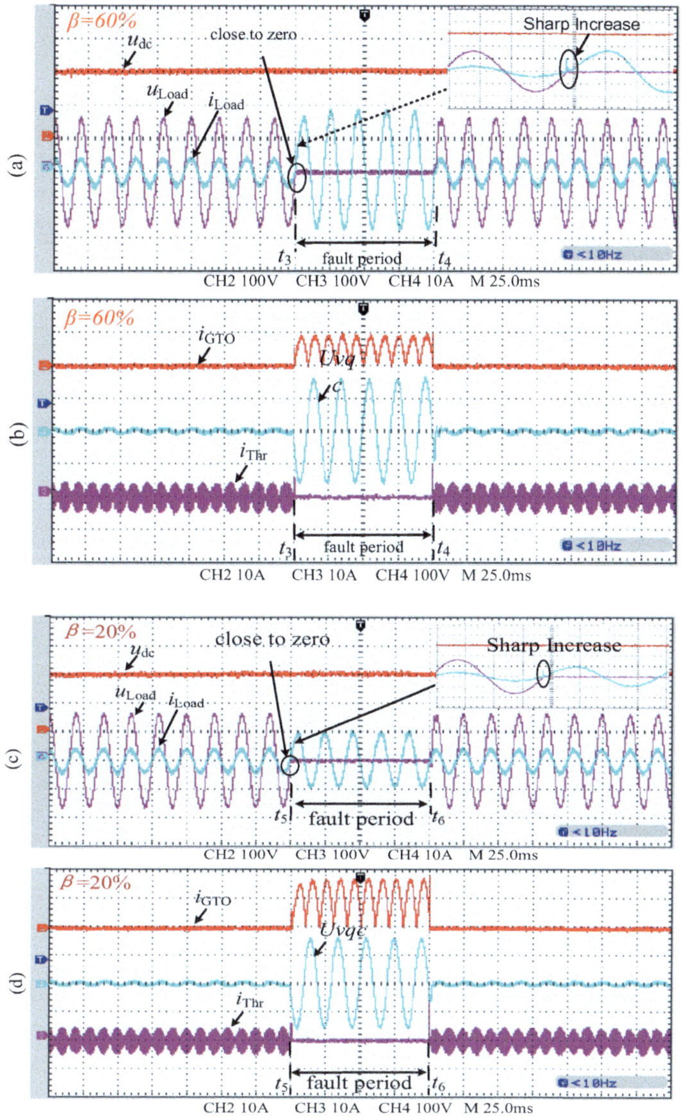

Fig. 5.14 Experimental results for the DF-VQC system with different duty cycles β when the fault causes the load current to go to zero

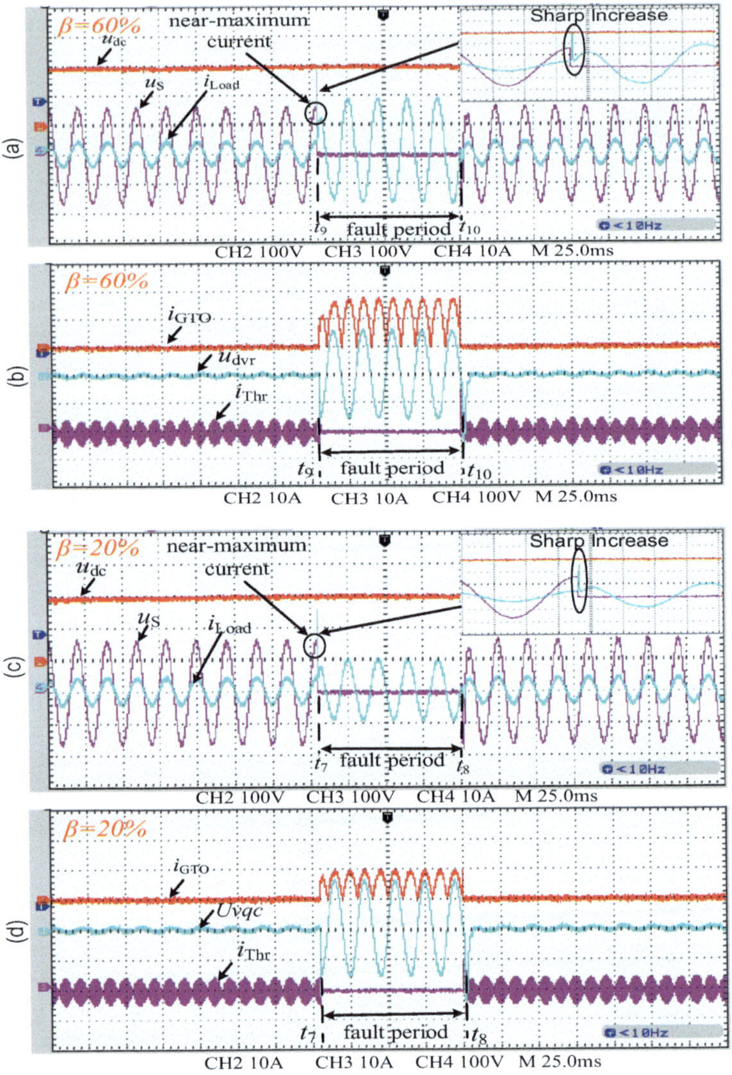

Fig. 5.15 Experimental results for the DF-VQC system with different duty cycles β when the fault causes a near-maximum current

References

1. T. Ghanbari, E. Farjah, Development of an efficient solid-state fault current limiter for microgrid. IEEE Trans. Power Del. **27**(4), 1829–1834 (2012)
2. W. Fei, Y. Zhang, Z. Meng, A novel solid-state bridge type FCL for three-phase three-wire power systems, in *Proceedings 22nd Annual IEEE applied Power Electronics Conference*, vol. 2, (2007), pp. 1084–1088

3. M. Tarafdar, M. Abapour, Nonsuperconducting fault current limiter with controlling the magnitudes of fault currents. IEEE Trans. Power Elect. **24**(3), 613–619 (2009)
4. T. Ise, N.H. Nguyen, S. Kumagai, Reduction of inductance and current rating of the coil and enhancement of fault current limiting capability of a rectifier type superconducting fault current limiter. IEEE Trans. Appl. Super. **11**(1), 1932–1935 (2001)
5. H.J. Boenig, D.A. Paice, Fault current limiter using a superconducting coil. IEEE Trans. Magn. **MAG-19**(3), 1051–1053 (1983)
6. S.B. Naderi, M. Jafari, M.T. Hagh, Controllable resistive type fault current limiter (CR-FCL) with frequency and pulse duty-cycle. Electr. Power Syst. Rese. **61**, 11–19 (2014)
7. Z. Lu, D. Jiang, Z. Wu, A new topology of fault-current limiter and its parameters optimization, in *Proceedings of IEEE Power Electronics Specialist Conference* (2003), pp. 462–465
8. M. Firouzi, G.B. Gharehpetian, M. Pishvaie, Proposed new structure for fault current limiting and power quality improving functions, in *Proceedings International Conference on Renewable and Power Quality (ICREPQ'10)*, Granada, Spain (2010), p. 1
9. H. Radmanesh, S.H. Fathi, G.B. Gharehpetian, A. Heidary, Bridge-type solid-state fault current limiter based on AC/DC reactor. IEEE Trans. Power Del. **31**(1), 200–209 (2016)
10. S.S. Choi, T.X. Wang, D.M. Vilathgamuwa, A series compensator with fault current limiting function. IEEE Trans. Power Del. **20**(3), 2248–2256 (2005)
11. M.R. Banaei, A. Nabavandi, S.H. Hosseini, Hysteresis controller based fault current interruption using DVR, in *Proceedings of Emerging Research Areas: Magnetics, Machines and Drivers (AICERA/iCMMD)*, Kottayam (2014), pp. 1–4
12. S. Zhikang, Y. Peng, Z.J. Shen, T. Chunming, J. Fei, C. Ying, Design considerations of a fault current limiting dynamic voltage restorer (FCL-DVR). IEEE Trans. Smart Grid **6**(1), 14–25 (2015)
13. F. Jiang, C. Tu, Z. Shuai, M. Cheng, Z. Lan, F. Xiao, Multilevel cascaded-type dynamic voltage restorer with fault current limiting function. IEEE Trans. Power Del. **31**(3), 1261–1269 (2016)
14. M.J. Newman, D.G. Holmes, J.G. Nielsen, F. Blaabjerg, A dynamic voltage restorer (DVR) with selective harmonic compensation at medium voltage level. IEEE Trans. Ind. Appl. **41**(6), 1744–1753 (2005)
15. S. Bacha, D. Frey, J.L. Schanen, E. Lepelleter, P.O. Jeannin, R. Caire, Short-circuit limitation thanks to a series connected VSC, in *Proceedings of Applied Power Electronics Conference and Exposition (APEC)*, Austin, TX (2008), pp. 1938–1945
16. N.H. Woodley, L. Morgan, A. Sundaram, Experience with an inverter-based dynamic voltage restorer. IEEE Trans. Power Del. **14**(3), 1181–1186 (1999)
17. A. Ghosh, G. Ledwich, Compensation of distribution system voltage using DVR. IEEE Trans. Power Del. **17**(4), 1030–1036 (2002)
18. D.A. Fernandes, F.F. Costa, M.A. Vitorino, A method for averting saturation from series transformers of dynamic voltage restorers. IEEE Trans. Power Delivery **29**(5), 2239–2247 (2014)
19. P. Jayaprakash, B. Singh, D.P. Kothari, A. Chandra, Al-Haddad K. Control of reduced-rating dynamic voltage restorer with a battery energy storage system. IEEE Trans. Ind. Appl. **50**(2), 1295–1303 (2014)
20. S. Gao, X. Lin, S. Ye, H. Lei, Y. Kang, Transformer inrush mitigation for dynamic voltage restorer using direct flux linkage control. IET Power Electron. **8**(1), 2281–2289 (2015)
21. S. Bernet, R. Teichmann, A. Zuckerberger, P.K. Steimer, Comparison of high-power IGBT's and hard-driven GTO's for high-power inverters. IEEE Trans. Indus. Appl. **35**(2), 478–482 (1999)
22. M.M.R. Ahmed, G. Putrus, L. Ran, L. Xiao, Harmonic analysis and improvement of a new solid-state fault current limiter. IEEE Trans. Ind. Appl. **40**(4), 1012–1019 (2004)

Part III
The Extended Application of Advanced Voltage Quality Controller

In this part, the application effect of Advanced Voltage Quality Controller in typical scenarios is introduced. For example, capacitively coupled VQC and DF-VQC have been extended for capacity optimization in UPQC and motor starting process, respectively. Then, the power rating of IUPQC's power electronic module can be significantly reduced, the overcurrent during the motor starting process can also be effectively reduced as well.

Chapter 6
Improved Single-Phase UPQC Based on Capacitively Coupled Voltage Quality Controller

This section introduces a new topology of UPQC. It has an LC filter coupled with an external capacitor at the series part. Besides, a comprehensive coordination strategy is proposed to operate the improved UPQC (IUPQC) at reactive power compensation (RPC) mode and voltage fluctuation compensation (VC). Specifically, to achieve the low power rating, the proposed control strategy makes the series part undertakes partial reactive power that needs to be compensated at the parallel side in RPC mode. At the same time, it makes the parallel side coordinates the series side to expand the optimal operation range in VC mode. Compared with the conventional UPQCs, the IUPQC can well reduce the power rating of back-to-back connected converters from both parallel and series sides. Moreover, the optimal parameters design method is also provided.

6.1 Introduction

The widespread adoption of power electronics and motor loads within electrical networks has introduced pressing concerns regarding power quality, primarily revolving around reduced power factors, harmonic contamination, and voltage waveform distortions, all of which can significantly degrade system performance and reliability [1, 2]. In response to these challenges, a variety of power quality enhancement devices have been developed and studied, including APFs, STATCOMs, and unified power quality controllers (UPQCs). Among these, the UPQC stands out as a comprehensive solution, as illustrated in Fig. 6.1 [3–6], due to its dual capability to address both voltage and current quality issues simultaneously [7, 8]. Recognizing its pivotal role in mitigating power quality problems, researchers have dedicated considerable efforts to refining control techniques [9–18] and exploring novel topologies

F. Jiang et al., *Advanced Voltage Quality Controller for New Power Distribution Systems*, https://doi.org/10.1007/978-981-97-9861-2_6

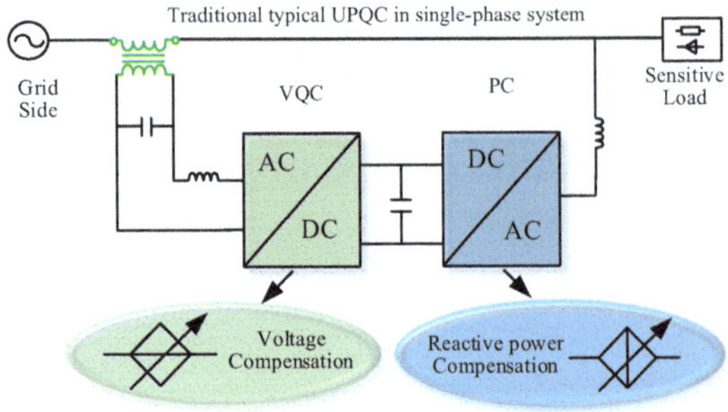

Fig. 6.1 Traditional typical configuration of UPQC in single phase

[19–22], reflecting the ongoing commitment to enhancing UPQC efficacy and versatility within the scientific community. For the governance of UPQCs, diverse control methodologies have been conceptualized and documented in academic literature. These strategies can be broadly categorized into UPQC-P [9, 10], UPQC-Q [11], and the UPQC- VA_{min} technique [14], which aims for minimal VA loading. However, the UPQC-P approach necessitates a higher power capacity for the parallel converter (PC), whereas the UPQC-Q methodology places similar demands on the VQC. The UPQC-VA_{min} method, while effective, can only mitigate the power requirements in VC mode.

To enhance efficiency and economize on the power ratings of both the series and parallel converters, coordinated control strategies have been proposed in [15–17]. These approaches seek to synchronize the operations of the series and parallel converters, aiming to optimize overall system performance. Nevertheless, as the severity of voltage sags escalates, the resultant surge in active power exchange between the series and parallel segments can lead to a rapid escalation in the requisite overall power rating.

Furthermore, an innovative data-driven control paradigm has emerged, incorporating variable phase angle control to ascertain the ideal power ratings for UPQCs [18]. While this strategy holds promise in minimizing the capacity requirements of UPQCs, its efficacy is somewhat curtailed by the limitations intrinsic to conventional topologies. This underscores the ongoing need for advanced topological designs and control methodologies that can further refine the performance and cost-effectiveness of UPQC implementations.

Regarding the topology of the UPQC, contemporary research predominantly revolves around enhancing the parallel side's efficiency and effectiveness. A novel configuration involving the insertion of a capacitor in series with the interface inductor on the parallel side has been suggested in [19, 20]. This additional capacitor facilitates the compensation of a portion of the load's reactive power, thereby diminishing the capacity requirements of the parallel converter and alleviating the demand for a high DC link voltage. Nonetheless, there remains room for further reduction in the power ratings of both the VQC and the PC during the voltage compensation phase. An UPQC design that integrates dual DC stages has been put forward [21], capable of achieving effective compensation at reduced DC link voltages. This topology not only boosts conversion efficiency but also markedly decreases the power rating of the DC-DC converter. Should the converter's power rating be further minimized during both reactive power compensation and voltage compensation stages, the overall capacity of the UPQC could be substantially reduced. A recent proposal [23] introduces a UPQC topology that incorporates an LC filter coupled with an external capacitor on the series side. This design boasts significant capacity optimization benefits, particularly evident at shallow sag depths. However, the full extent of its advantages under varying sag depths and power factors necessitates further exploration. Moreover, the systematic and comprehensive investigation into the parameter design for both the PC and VQC components remains an open area of research, requiring deeper scrutiny to fully leverage the potential of such topologies.

To address the limitations inherent in conventional UPQC designs, this chapter embarks on a comprehensive examination of the improved UPQC (IUPQC) output attributes under diverse operational scenarios. It delves into an optimized operational strategy and a refined parameter design procedure to accommodate a spectrum of working conditions. The principal contributions encapsulated in this chapter can be delineated as follows:

(1) Enhanced Topology and Dual Optimization: The IUPQC, featuring an LC filter coupled with an external capacitor on the series side, is subjected to an in-depth study. This configuration ensures concurrent optimization of the power ratings for both the VQC and the PC during voltage fluctuation compensation (VC) and reactive power compensation (RPC) modes, respectively.
(2) Coordinated Operation Strategy: An integrative coordination strategy is devised for the IUPQC, enabling the series component to share the reactive power load with the parallel segment during RPC mode. Conversely, the parallel part effectively extends the optimal operational range of the series component in VC mode, thereby enhancing the overall effectiveness and flexibility of the UPQC system.
(3) Comparative Analysis: The voltage and active/reactive power characteristics of the IUPQC are meticulously analyzed, offering a transparent perspective on its superiority with conventional UPQCs employing traditional control strategies and modified schemes on the parallel side.
(4) Parameter Design Methodology: A systematic approach to parameter design is proposed, facilitating the harmonization of the series and parallel sections under varying voltage sag depths and load power factors. This ensures the IUPQC's adaptability and optimal performance across a wide range of operational conditions.

6.2 Topology of Proposed IUPQC

Figure 6.2 illustrates the architecture of the IUPQC. The series portion is interfaced with the power source, while the parallel section is connected to the load. Both sections are interconnected in a back-to-back configuration, sharing a common DC

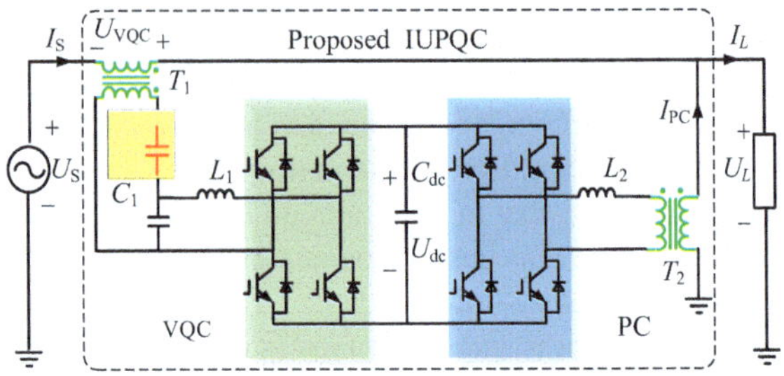

Fig. 6.2 Topology of the improved UPQC

bus capacitor C_{dc}. Beyond the series converter, the series section incorporates an *LC* filter, a series transformer T_1, and a series coupling capacitor C_S [23], enhancing its functionality. The parallel section, on the other hand, comprises a parallel converter, an *L* filter, and a parallel transformer T_2. In Fig. 6.2, U_S, U_L, and U_{VQC} represent the source voltage, load voltage, and compensation voltage, respectively. U_{dc} denotes the DC link voltage, while I_S, I_L, and IPC signify the source current, load current, and PC compensated current, respectively.

In the normal condition, the voltage phasor diagram of VQC in traditional UPQC [6] is elaborated in Fig. 6.3a, where U_S is considered as the reference phasor, U_S', U_L' and I_L' represent the grid voltage, load voltage, and current after the VQC activation.

As shown in Fig. 6.3a, the reactive power outputted by VQC is

$$Q_{VQC} = (U_L I_L \sin\theta_L)/2 - (U_S I_L \sin(\theta_L - \alpha))/2, \tag{6.1}$$

Where θ_L is the load power factor angle, α is the phase difference between U_S and U_L after VQC activation, and σ is the phase angle jump of U_S after sag occurs [14]. Based on (6.1), it can be obtained that the reactive power outputted by VQC is proportional to angle α. Thus, as long as the phase angle of VQC outputted voltage is reasonable, VQC can share reactive power with the PC.

In the voltage sag condition, the voltage phasor diagram of VQC in traditional UPQC is elaborated in Fig. 6.3b [22]. The active power and voltage outputted by VQC is

$$\begin{aligned} P_{VQC} &= (U_L I_L \cos\theta_L)/2 - (U_S' I_L \cos(\theta_L - \alpha))/2 \\ U_S' &= (1 - k_{sag}) U_L, \end{aligned} \tag{6.2}$$

where k_{sag} is defined as the voltage drop depth and is equal to $(1\text{-}U_S')/\,U_S$. Based on (6.2), it can be obtained that the active power output by VQC can be controlled to 0 while $1 - k_{sag} > \cos\theta_L$; once $1 - k_{sag} < \cos\theta_L$, the P_{VQC} is greater than zero, and by increasing the value of α, the active power required for VQC output can be decreased.

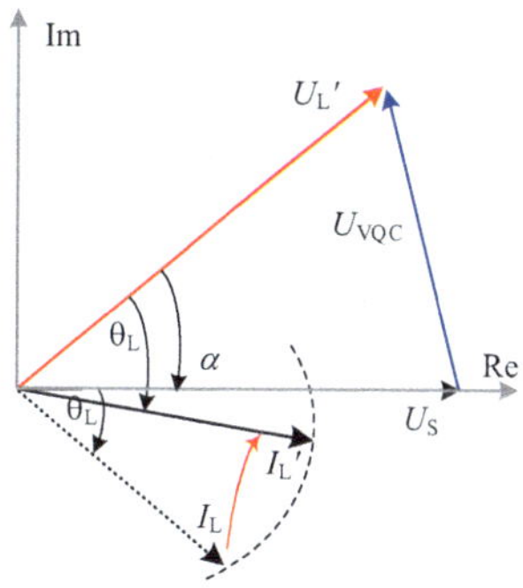

(a) In the normal condition

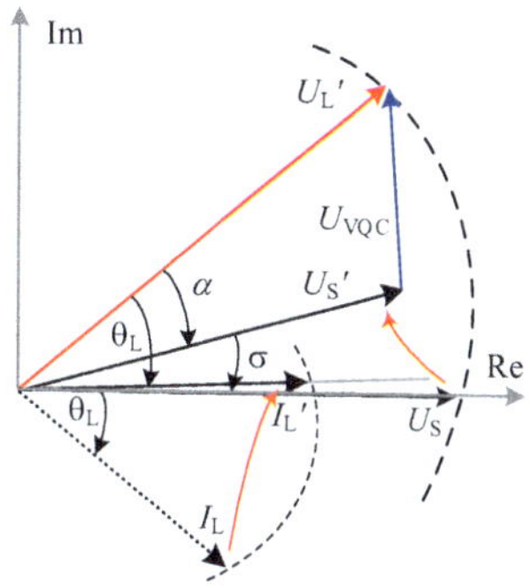

(b) In the voltage sag condition

Fig. 6.3 Phasor diagrams of VQC

Further, the expression of VQC output voltage during two conditions can be expressed as follows

$$U_{VQC} = \sqrt{U_L^2 + U_S^2 + 2U_L U_S \cos\alpha}. \tag{6.3}$$

Given the conditions stipulated by Eqs. (6.1)–(6.3), if the value of L' (representing a specific inductance parameter) is less than that of L, an augmentation in the magnitude of L' results in an enhanced reactive power sharing capability for the series portion during RPC mode. Concurrently, during voltage control mode, this increase minimizes the active power exchange between the series and parallel components, subsequently lessening the capacity requirements for the parallel section. Nevertheless, this improvement is accompanied by an elevated output voltage from the VQC, which subsequently gives rise to a greater demand for overall system capacity.

6.3 Operation Principle of the Proposed IUPQC

In this segment, we delve into the operational dynamics and efficacy of the proposed IUPQC in both RPC mode and VC mode. Our theoretical exploration encompasses the harmonized functioning of the VQC and PC, elucidating the underlying mechanisms that enable capacity optimization. We also outline the methodology for establishing the reference values for output voltage and current in the VQC and PC components.

Moreover, a comparative assessment is provided, juxtaposing the performance of our proposed IUPQC against conventional strategies to highlight its superiority in terms of efficiency and adaptability. Throughout this manuscript, the load voltage and current serve as pivotal reference points, guiding the analysis and evaluation of the IUPQC's operational capabilities in managing power quality under varying conditions.

This comprehensive analysis is intended to substantiate the IUPQC's potential for enhancing power quality by effectively addressing both reactive power and voltage fluctuation challenges, thereby contributing to the advancement of power distribution systems. By focusing on the coordinated operation principles and capacity optimization mechanisms, we aim to offer insights into how the IUPQC can revolutionize power quality control, making it a compelling alternative to traditional power quality enhancement techniques.

6.3.1 Reactive Power Compensation Mode in Normal Condition

The operational principle and voltage phasor representation of the IUPQC operating in RPC mode are detailed in Fig. 6.4. In this mode, the series and parallel sections engage in power transactions with the system by injecting voltage and current, respectively. For the series section, the series coupling capacitor (C_S) significantly contributes to the required output voltage (U_{VQC}), with the voltage vector U_{VQC} being the sum of the voltage vectors U_{conv} and U_{CS}. Concurrently, the parallel section handles both reactive power output and the essential active power exchange with the series section. Throughout the reactive power compensation process, the magnitudes of U_L and I_L remain stable, and the phase angle of U_S aligns perfectly with I'_S, as illustrated in Fig. 6.4.

The power flow in RPC mode is shown in Fig. 6.5. The load-reactive power compensation is jointly undertaken by the series part and parallel part. A large part of Q_L is provided by the series coupling capacitor C_S defined as Q_{CS1}, and the rest is output by the VQC and PC, defined as Q_{conv1} and Q_{PC1}, respectively. Then, from Fig. 6.4, we can get the expression of Q_{VQC1} shown as

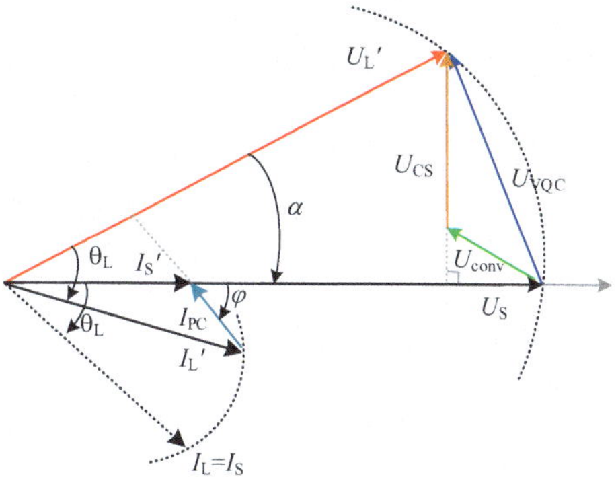

Fig. 6.4 Phasor diagrams during the normal stage using the proposed IUPQC

$$Q_{VQC1} = Q_{CS1} + Q_{conv1} = U_{VQC} I_S' \sin\left(\frac{\pi - \alpha}{2}\right) = U_{VQC} I_S' \cos(\alpha/2). \tag{6.4}$$

Considering that the output active power of the grid side is constant during the RPC mode, it refers to $P_S' = P_L$, so we have

$$I_S' = I_L \cos(\theta_L). \tag{6.5}$$

Then, based on the relation between U_L, U_{VQC}, and U_S', the magnitude and the phase angle of U_{VQC} are expressed as follows

$$\begin{aligned} |U_{VQC}|^2 &= |U_L'|^2 + |U_S|^2 - 2|U_S'||U_L'|\cos(\alpha) \\ &= 2(1 - \cos(\alpha)). \end{aligned} \tag{6.6}$$

Based on (6.4) and (6.6), α can be obtained while the proportion of reactive power compensation shared by the series part is determined

$$\alpha = \arccos\left(\sqrt{1 - \left(Q_{VQC1}/\cos(\theta_L)\right)^2}\right), \tag{6.7}$$

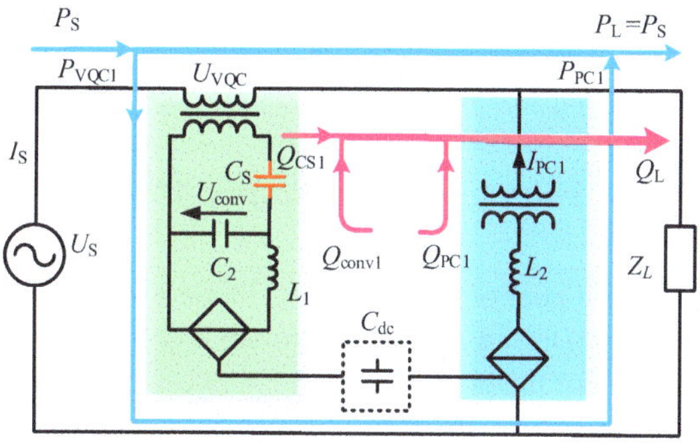

Fig. 6.5 Power flow during the normal stage using the proposed strategy

where the amplitudes of U_L', U_S', U_S, and I_L are equal to 1 p.u in RPC mode. Further, due to the active power absorbed by the series part being the same as that active power outputted by the parallel part in the RPC mode, we can get that

$$\begin{aligned} P_{\mathrm{PC1}} &= P_{\mathrm{VQC1}} = U_{\mathrm{VQC}} I_S' \cos((\pi - \alpha)/2) \\ &= \cos(\theta_L)\left(1 - \sqrt{1 - \left(Q_{\mathrm{VQ1}}/\cos(\theta_L)\right)^2}\right), \end{aligned} \tag{6.8}$$

$$S_{\mathrm{PC1}} = I_{\mathrm{PC1}} = \sqrt{\left(P_{\mathrm{VQC1}}\right)^2 + \left(Q_L - Q_{\mathrm{VQC1}}\right)^2}, \tag{6.9}$$

where $P_{\mathrm{PC1}}, P_{\mathrm{VQC1}}$ are the outputted active power by the parallel part and series part, respectively. S_{PC1} is the apparent power of the parallel part.

Then, based on the relation among I_{PC1}, I_S', and I_L', the phase angle difference between I_{PC} and U_S is expressed as follows

$$\varphi = -\arccos\left(\left(\left|I_S'\right|^2 + |I_{\mathrm{PC1}}|^2 - |I_L|^2\right) / 2|I_{\mathrm{PC1}}|\left|I_S'\right|\right). \tag{6.10}$$

After the coordination between VQC and series coupling capacitor C_S, the apparent power of VQC (S_{conv}) and U_{conv} can be depicted as

$$S_{\mathrm{conv1}} = \sqrt{(P_{\mathrm{VQC1}})^2 + \left(Q_{\mathrm{VQC1}} - |I_S'|^2(1/\omega C_S)\right)^2}, \tag{6.11}$$

$$|U_{\mathrm{conv1}}| = S_{\mathrm{conv1}} / \left|I_S'\right| = S_{\mathrm{conv1}} / \cos(\theta_L). \tag{6.12}$$

Then, the curves of U_{VQC}, U_{conv}, S_{conv}, and Q_L are shown in Fig. 6.6 when the load-reactive power shared on the series part increased from 0.3 to 0.7, the value of X_{CS} is 0.5 p.u, and the load power factor is 0.7. As shown in Fig. 6.6, the power rating of the parallel part decreases gradually with the increase of load-reactive power shared by VQC. Further, as can be seen from the red and green lines, the voltage rating of VQC is almost half of the whole series part (U_{VQC} can represent the voltage rating of VQC in traditional UPQC with UPQC-S strategy). Thus, the proposed IUPQC not only improves the utilization rate of VQC, but also reduces the overall power rating demand.

6.3.2 Voltage Compensation Mode in Voltage Sag Condition

The main goal of IUPQC in VC mode is to keep the amplitude of load-side voltage constant. During this period, the smaller the active power required by the series part, the smaller the power exchange between VQC and PC.

Considering the condition that no power exchange between VQC and PC can effectively reduce the power level requirements of both. As for the case that $1 - k_{\mathrm{sag}}$

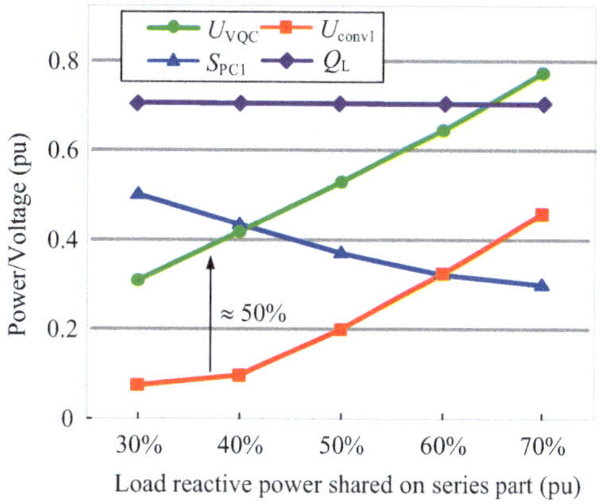

Fig. 6.6 Voltage and power rating of IUPQC in RPC mode

$< \cos\theta_L$, the optimal operation point can be achieved by adjusting the power factor angle downstream of VQC through the PC based on (6.2). This section focuses on the analysis of the energy flow relation between VQC and PC during the coordination period and discusses the optimal operation parameters of IUPQC in voltage sag conditions.

The operation principle and voltage phasor diagram of the IUPQC in VC mode is elaborated in Fig. 6.7. In this case, the parallel part absorbing inductive reactive power as can be seen from I_{PC}, and through the superposition fundamental components of U_{conv}, U_{CS}, and U_S', the amplitude of load voltage U_L can be kept constant. Further, due to the U_{VQC} and U_{CS} being perpendicular to I_S', the phase angle of U_{conv} will be in the same phase as U_{CS} based on the Kirchhoff's voltage law (KVL).

Again, assuming that the output active power of the grid side is constant during the VC mode, it refers to $P_S' = P_L$, so we have

$$\begin{aligned} |I_S'||U_S'| &= U_L I_L \cos(\theta_L) \\ I_S' &= \cos(\theta_L) / (1 - k_{sag}), \end{aligned} \tag{6.13}$$

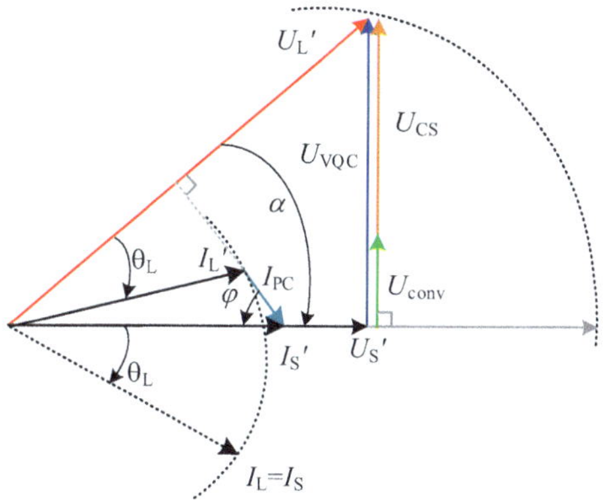

Fig. 6.7 Phasor diagrams during the voltage compensation stage using the proposed IUPQC

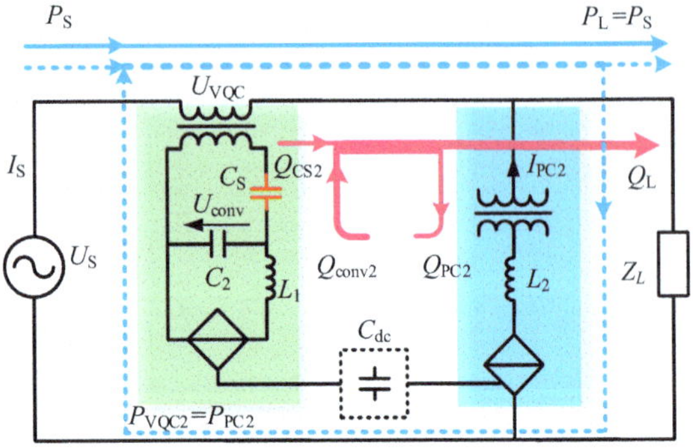

Fig. 6.8 Power flow during the voltage compensation stage using the proposed strategy

where the amplitude of U_L' and U_S' is equal to 1 p.u and $(1 - k_{sag})$ p.u in VC mode. As seen in Fig. 6.7, we have $U_L' \cos(\alpha) = U_S'$, and the magnitude and the phase angle of U_{VQC} are expressed as follows

$$\cos(\alpha) = 1 - k_{sag}, \tag{6.14}$$

$$U_{VQC} = \left|U_L'\right| \sin(\alpha) = \sqrt{2k_{sag} - \left(k_{sag}\right)^2}. \tag{6.15}$$

Further, based on (6.13), (6.15), and the reactive power flow in VC mode in Fig. 6.8, we can get the expression of Q_{VQC2} and Q_{PC2} shown as

$$Q_{VQC2} = Q_{PC2} + Q_L = U_{VQC} I_S' = \sqrt{2k_{sag} - (k_{sag})^2} \frac{\cos(\theta_L)}{1 - k_{sag}}, \tag{6.16}$$

$$\begin{aligned} Q_{PC2} &= \sqrt{2k_{sag} - (k_{sag})^2}\left(\cos(\theta_L)/\left(1 - k_{sag}\right)\right) - Q_L \\ I_{PC2} &= Q_{PC2}, \end{aligned} \tag{6.17}$$

where the variables are calculated in the form of unit values. Then, based on the relation between I_{PC}, I_S', and I_L', the phase angle difference between I_{PC} and U_S is expressed as follows

$$\varphi = (\pi/2) - \alpha = (\pi/2) - \arccos(1 - k_{sag}). \tag{6.18}$$

After the coordination between VQC and series coupling capacitor C_S, the apparent power of VQC (S_{conv2}) and U_{conv} can be depicted as

$$S_{conv2} = |U_{conv}|\left|I_S'\right| = \left(\left|U_{VQC}\right| - \left|I_S'\right|(1/\omega C_S)\right)\left|I_S'\right|, \tag{6.19}$$

$$|U_{conv}| = S_{conv2}/\left|I_S'\right|. \tag{6.20}$$

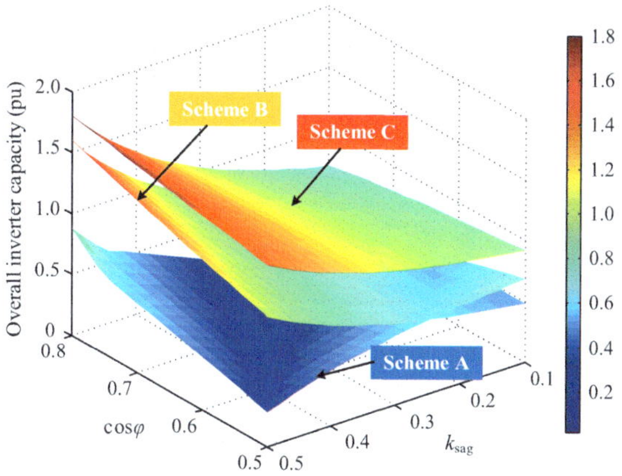

Fig. 6.9 Overall converter capacity of three scenarios under different working conditions

Then, the S_{UPQC2} can be expressed as

$$S_{UPQC2} = Q_{PC2} + S_{conv2} = 2\sqrt{2k_{sag} - (k_{sag})^2}\frac{\cos(\theta_L)}{1 - k_{sag}} - Q_L - I'_S\frac{1}{\omega C_S}. \quad (6.21)$$

The calculation method of reference value during the voltage swell stage is similar to the case of voltage sag described in this chapter, so it is not repeated here.

Based on the above analysis, the overall converter rating comparison is shown in Fig. 6.9 among the proposed IUPQC (scheme A), the improved scheme on the parallel side of UPQC [15] (scheme B), and the traditional topology with UPQC-P strategy (scheme C). The rating of scheme A is significantly lower than that of schemes B and C. Note that, if only to compensate for such slight voltage sags, the capacity can be further reduced by increasing the reactance of C_S.

Further, it is also worth noting that when the parallel part outputs only inductive reactive power, the capacity of the parallel part increases rapidly with the increase of the sag depth while $1 - k_{sag} < \cos\varphi$, as can be seen from the blue line in Fig. 6.10. Therefore, in order to realize the overall capacity optimization of IUPQC in VC mode, the reactive power of the PC output needs to be designed reasonably. Thus, it is assumed that Q_{PC2} does not exceed the optimized capacity requirements of PC in RPC mode shown as

$$Q'_{PC2} = kS_{PC1}(0 < k < 1). \quad (6.22)$$

Then, the Q_{VQC2}, the magnitude and the phase angle of U_{VQC} are expressed as follows

$$Q'_{VQC2} = Q'_{PC2} + Q_L = U'_L I'_S \sin(\alpha'), \quad (6.23)$$

$$\sin(\alpha') = \left(Q'_{PC2} + Q_L\right)(1 - k_{sag})/\cos(\theta_L), \quad (6.24)$$

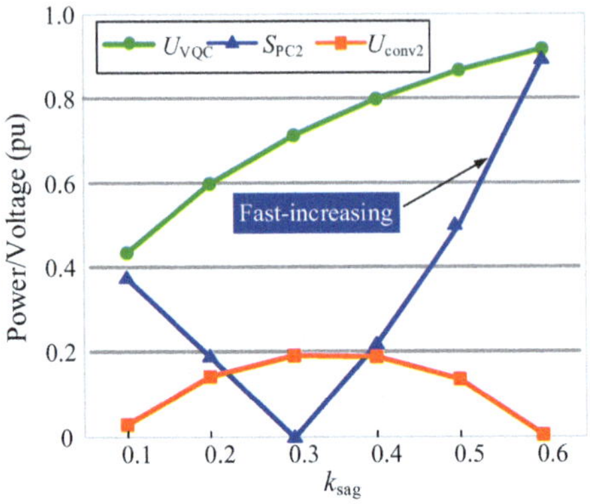

Fig. 6.10 Voltage and power rating of IUPQC in VC mode

$$U_{\text{VQC}} = \left|U_{\text{L}}'\right|^2 + \left|U_{\text{S}}'\right|^2 - 2\left|U_{\text{L}}'\right|\left|U_{\text{S}}'\right|\cos(\alpha\prime). \tag{6.25}$$

Further, the active power outputted by the VQC is the same as that absorbed active power by the PC in the VC mode, so we can get that

$$S_{\text{PC2}}' = P_{\text{VQC2}} = U_{\text{L}}' I_{\text{S}}' \cos\left(\alpha'\right) - P_{\text{L}}. \tag{6.26}$$

Then, the amplitude of U_{L}' remains 1p.u after IUPQC is activated in VC mode, the apparent power and compensated current of PC in p.u value is expressed as

$$S_{\text{PC2}}' = U_{\text{L}}' I_{\text{PC2}}' = \sqrt{\left(P_{\text{VQC2}}'\right)^2 + \left(Q_{\text{PC2}}'\right)^2}, \tag{6.27}$$

$$cos\left(\pi - \varphi'\right) = \left(\left|I_{\text{S}}'\right|^2 + |I_{\text{PC2}}|^2 - |I_{\text{L}}|^2\right) \Big/ \left(2|I_{\text{PC2}}|\left|I_{\text{S}}'\right|\right). \tag{6.28}$$

In this condition, the S_{conv2} and U_{conv} can be depicted as

$$S_{\text{conv2}}' = \sqrt{(P_{\text{VQC2}})^2 + \left(Q_{\text{VQC2}} - \left|I_{\text{S}}'\right|^2 (1/\omega C_{\text{S}})\right)^2}, \tag{6.29}$$

$$|U_{\text{conv}}|' = S_{\text{conv2}}' \big/ \left|I_{\text{S}}'\right|. \tag{6.30}$$

Then, based on (6.22) and (6.23), the apparent power rating optimization of the IUPQC in VC can be formulated in the following

$$S_{2(\min)} = \min(S_{\text{UPQC2}}') = \min(S_{\text{PC2}}' + S_{\text{conv2}}'). \tag{6.31}$$

By solving the minimum value of S_{UPQC2}' under inequality constraints (6.22), the optimal operation parameters of IUPQC in VC mode are confirmed.

6.4 Power Rating Selection and Overall Control of Proposed IUPQC

The proposed IUPQC may operate in different modes, such as VC mode and RPC mode, according to the state of the grid. In this section, the optimal design of system capacity to meet both functional requirements is carried out in this section, as well as includes the overall control strategy of IUPQC.

6.4.1 Power Rating Selection of Proposed IUPQC

(1) Design of external capacitor C_S: assumed the rated apparent capacity of load is S_L, the reactive power shared on VQC at least more than $m_1{*}Q_L$ and the reactive power shared on C_S at least more than $m_2{*}Q_{VQC}$, where $0 < m_1$, $m_2 < 1$.

Considering the rated load current and Q_{CS} are derived from (6.32) and (6.33)

$$I_L = S_L/U_S, \tag{6.32}$$

$$Q_{C_S} = (I_S)^2/\omega C_S. \tag{6.33}$$

Then, based on (6.9) and (6.11), the apparent power rating optimization of the IUPQC in RPC mode can be formulated in the following

$$S_{1(\min)} = \min(S_{PC1} + S_{conv1}), \tag{6.34}$$

subject to

$$(m_1 * Q_L) \le Q_{VQC} \le Q_L, \tag{6.35}$$

$$\left((S_L/U_S)^2 \Big/ Q_{VQC}\right) \le C_S \le \left((S_L/U_S)^2 \Big/ (m_2 Q_{VQC})\right). \tag{6.36}$$

By solving the minimum value of S_{UPQC1} under inequality constraints (6.35) and (6.36), the optimal value of coupling capacitance is confirmed.

(2) Design of Power Converter: the values of S_{PC1} and S_{conv1} can be derived after the C_S is calculated based on (6.9) and (6.11). Further, it is assumed that the proposed IUPQC system has a compensation ability to cover the sag issue with a sag depth up to k_{max}. Then, the optimal value of k is confirmed based on (6.31). Thus, the value of S'_{PC2} and S'_{conv2} can be derived as well.

Finally, the capacities of VQC and PC should meet the following requirements

$$S_{PC} = \max(S_{PC1}, S'_{PC2})$$
$$S_{conv} = \max(S_{conv1}, S'_{conv2}). \tag{6.37}$$

(3) DC Link Voltage Selection: the amplitudes of system current in RPC and VC mode are $I_L\cos\theta_L$ and $I_L\cos\theta_L/(1 - k_{sag})$, respectively. The rated output voltage and DC link voltage of VQC can be obtained as

$$U_{conv} = S_{conv}/I_{Smin} = \max\left(\frac{S_{conv1}}{I_L\cos\theta_L}, \frac{S'_{conv2}(1 - k_{sag})}{I_L\cos\theta_L}\right), \tag{6.38}$$

$$U_{dc} = U_{conv}/m = \max\left(\frac{S_{inv1}}{mI_L\cos\theta_L}, \frac{S'_{conv2}(1 - k_{sag})}{mI_L\cos\theta_L}\right), \tag{6.39}$$

where m is the modulation ratio of the VQC in IUPQC.

Moreover, the selection of the series transformer's power rating ought to be approached with consideration for minimizing the aggregate expenditure associated with the transformer itself, along with the series and parallel converters. Given that the filter parameters for both the series and parallel sections are akin to those found in traditional UPQC configurations, they will not be exhaustively discussed herein.

To guarantee that the requisite DC link voltage for the series converter aligns with that of the parallel converter, the turns ratio of the parallel transformer should be configured to exceed a 1:1 ratio. This strategic adjustment serves to decrease the operational voltage rating required by the parallel converter, thus potentially reducing costs and enhancing efficiency.

This matter has been comprehensively explored in previous literature [19, 20], where the inclusion of a coupling capacitor on the parallel side obviates the need for a parallel transformer. This innovative design not only streamlines the system architecture but also contributes to a more economical and efficient power quality solution. By eliminating the parallel transformer, the system can achieve a more compact design while potentially reducing losses and improving overall system performance.

6.4.2 Overall Control Strategy of Proposed IUPQC

Figure 6.11 illustrates the block diagram of the proposed scheme, using phase A as a representative example. During normal grid voltage conditions, the UPQC operates in RPC mode to enhance the power factor on the system side. However, when grid voltage fluctuations occur, the UPQC switches to VC mode, maintaining a constant voltage amplitude at the load side while simultaneously correcting the power factor on the system side. If the sag factor (k_{sag}) exceeds the reference value (k_{ref}) [18], it

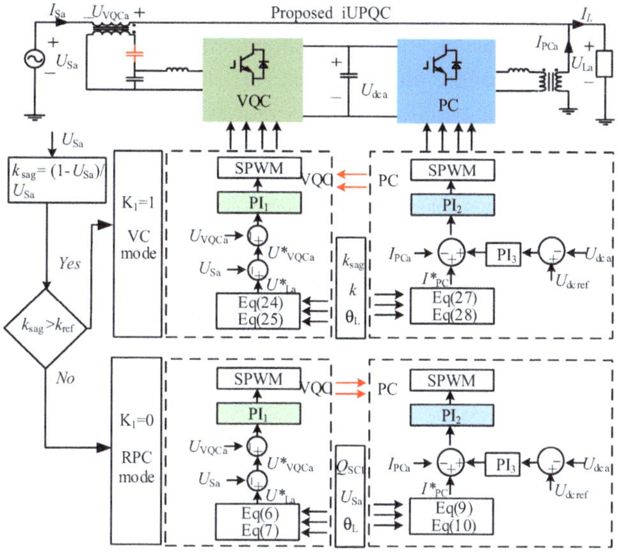

Fig. 6.11 Overall control scheme of proposed IUPQC

indicates the occurrence of a sag event, triggering the logical judgment K_1 to equal 1. When K_1 equals 0, the IUPQC operates in RPC mode. Initially, specific values are derived from Eqs. (6.6) and (6.7). Subsequently, I_{PCf} and the corresponding value are calculated using Eqs. (6.9) and (6.10). Finally, the reference voltage for the VQC (U_{VQCa}) and the reference current for the parallel converter (I_{PCa}) are generated.

When k_{sag} exceeds k_{ref}, the VC mode is activated. Subsequently, the reference values for the VQC injected voltage and PC injected current are derived using the same methodology outlined earlier, based on Eqs. (6.24), (6.25), (6.27), and (6.28). Irrespective of the mode in which the IUPQC operates, a PI controller is employed to precisely track the reference values U_{VQCa} and I_{PCa}, as depicted in Fig. 6.12. In scenarios where the IUPQC encounters nonlinear load currents or distorted system voltage conditions, the voltage at the series transformer side serves as the feedback variable for the voltage control loop.

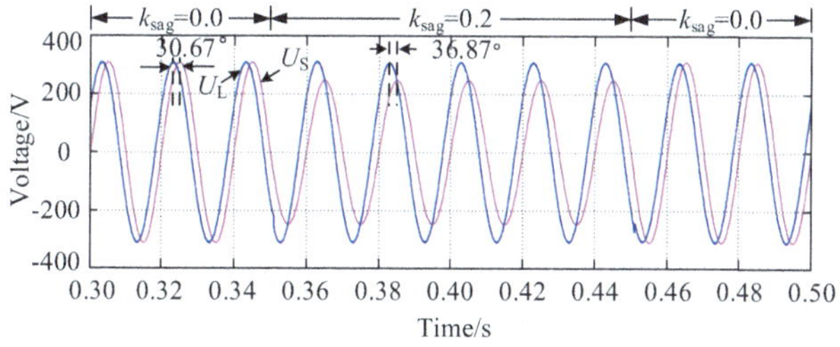

Fig. 6.12 Simulation waveforms of proposed IUPQC in U_S, U_L

Consequently, the converter section will generate a counter-harmonic voltage for compensation purposes, which effectively mitigates the harmonic constituents within the load voltage. Furthermore, the harmonic current elements on the load side are entirely neutralized by the parallel segment of the IUPQC.

Given that the central focus of this chapter is centered on optimizing the capacity of the IUPQC rather than introducing novel modulation techniques, fault identification methods, harmonic component extraction algorithms, or advanced controller designs for enhanced performance, these specific areas are not extensively covered. For comprehensive insights into these topics, readers are encouraged to consult relevant literature sources [24–30], where detailed discussions and analyses are provided.

6.5 Simulation Results

To substantiate the practicality and efficacy of the proposed IUPQC and its operational tactics, a simulation model for a single-phase IUPQC has been constructed within the MATLAB/Simulink environment for evaluative analysis. The pertinent parameters for this model are delineated in Table 6.1.The system model's performance is assessed utilizing the ode23tb solver, which is embedded within a variable-step simulation framework, with a sampling interval set at 20 microseconds. The switching frequency applied to the IGBT-based voltage source converter (VSC) is maintained at 10 kilohertz throughout the simulation process.

Table 6.1 Parameters used in the simulation

Parameters	Value
System voltage amplitude and frequency	310 V, 50 Hz
Load parameters R_L and L_L	7.7 Ω, 25 mH
DC link capacitor C_{dc}	4000 μF
DC link voltage of IUPQC U_{dc}	200 V
LC filter inductor in VQC L_1	2 mH
LC filter capacitor in VQC C_1	50 μF
LC filter inductor in PC L_2	4 mH
Series coupled capacitor C_S	0.5mF
Switching frequency	10 kHz
$k_{p1}, k_{i1}, k_{p2}, k_{i2}, k_{p3}, k_{i3}$	5; 0.5; 1.2; 0.5; 10; 0.5
Series transformer turn ratio	1:1
Parallel transformer turn ratio	2:1

6.5.1 Performances of Proposed IUPQC During 20% Voltage Sag Stage with Linear Load

A voltage sag with a 20% depth is introduced during the interval of 0.35–0.45 s, while the remaining time frame represents the normal steady-state condition. Consequently, the IUPQC operates in RPC mode during 0.30–0.35 s and 0.45–0.50 s. During these periods, the amplitude between U_L and U_S remains identical, and a phase difference of 30.67 degrees is maintained through the coordinated action of the VQC and PC. Notably, Figs. 6.12 and 6.13 clearly show that U_S is in phase with I_S. Furthermore, the required output compensation current I_{PC} of the parallel side is reduced from 20.2A to 10.29A due to the series side contributing half of the load's reactive power, as evident in Fig. 6.14. Despite U_{VQC} being significant after sharing the reactive power, similar to UPQC-Q (164.5 V, 0.53 p.u.), the presence of the coupling capacitor ensures that the amplitude of U_{conv} remains low (49.7 V, 0.16 p.u.), as demonstrated in Figs. 6.15 and 6.16. Therefore, the proposed IUPQC with its coordination strategy effectively reduces the capacity requirements for both the series and parallel sides simultaneously (Fig. 6.17).

During the voltage compensation phase, given that the sag depth meets the criterion $1 - k_{sag} > \cos\beta$, the operating parameters α and β can be computed as 36.87 and 53.12, respectively, based on Eqs. (6.14) and (6.18). As depicted in Fig. 6.12, the amplitude of U_L remains unaffected by the grid voltage sag due to the coordinated action between the VQC and PC. Despite the significant U_{VQC} value (186.4 V, 0.6 p.u.), the amplitude of U_{conv} remains low (28.3 V, 0.093 p.u.) because the coupling

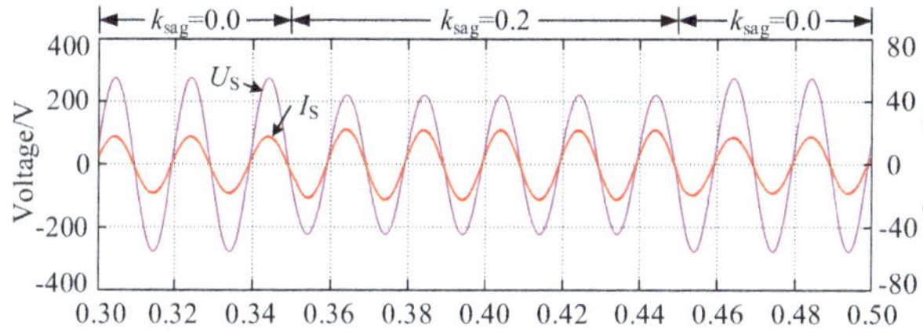

Fig. 6.13 Simulation waveforms of proposed IUPQC in U_S, I_S

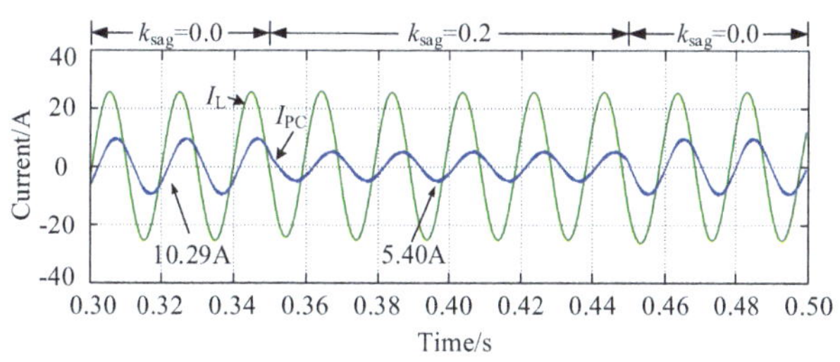

Fig. 6.14 Simulation waveforms of proposed IUPQC in I_L, I_{PC}

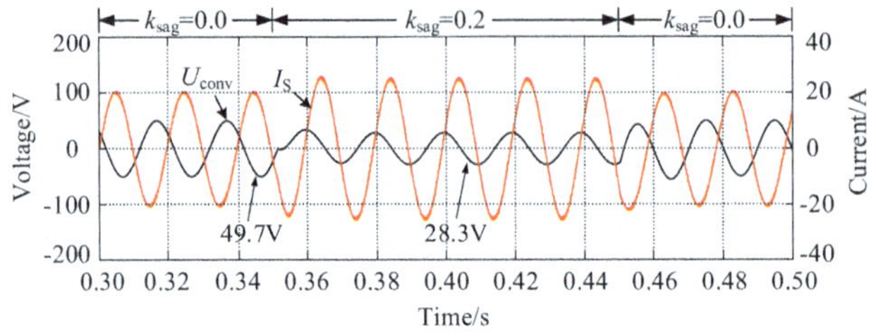

Fig. 6.15 Simulation waveforms of proposed IUPQC in U_{conv}, I_S

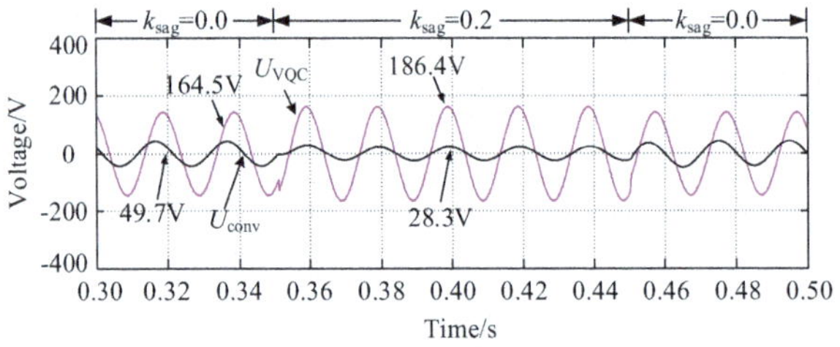

Fig. 6.16 Simulation waveforms of proposed IUPQC in U_{conv}, U_{VQC}

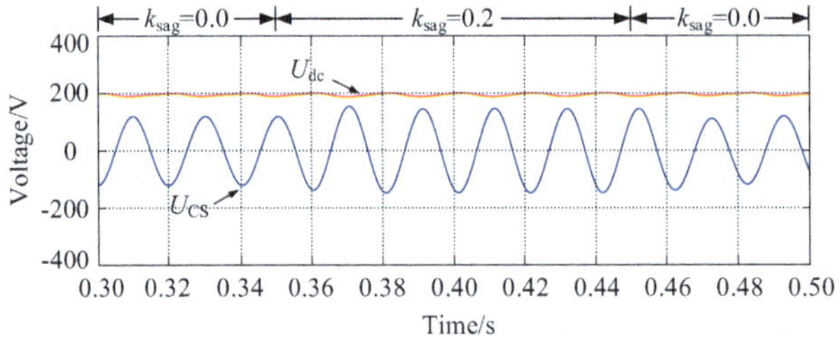

Fig. 6.17 Simulation waveforms of proposed IUPQC in U_{dc}, U_{CS}

capacitor shares a substantial portion of the compensated voltage, as evident in Figs. 6.15 and 6.16.Furthermore, the increase in the amplitude of I_S, as shown in Fig. 6.14, leads to an increase in the reactive power output by the coupling capacitor, thereby reducing the reactive power required from the parallel part to a smaller value (0.19 p.u.). Additionally, it is evident from Figs. 6.18 and 6.19 that P_S equals P_L, and Q_L is the sum of Q_{VQC} and Q_{PC}, confirming the reactive power compensation capability of the IUPQC and the efficacy of the coordination between the series and parallel sides.

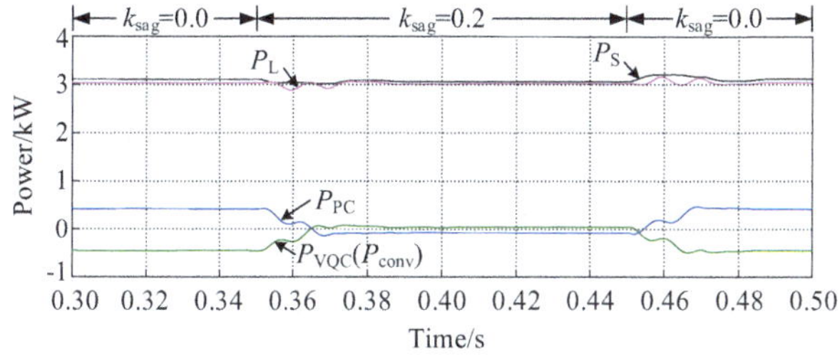

Fig. 6.18 Active power waveforms of proposed IUPQC

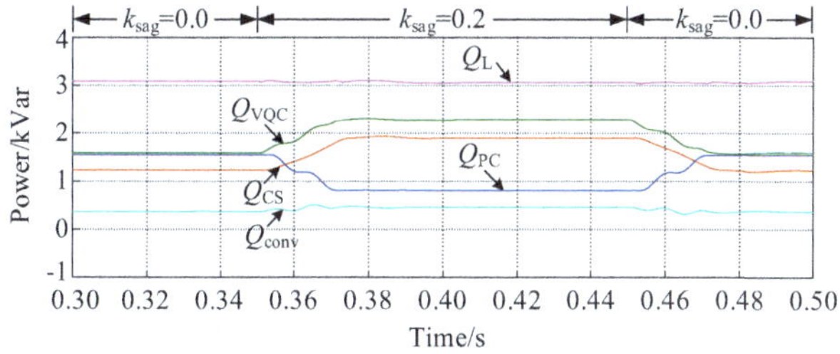

Fig. 6.19 Reactive power waveforms of proposed IUPQC

6.5.2 *Performances of Proposed IUPQC Under 50% Voltage Sag Case*

A voltage sag case with a depth of 50% is introduced during 0.35–0.45 s. As can be seen in Fig. 6.20, the amplitude of U_L is not affected. Although the U_{VQC} is very large (269.3 V, 0.866 p.u), the amplitude of U_{conv} still remains at a small value (17.6 V, 0.057p.u), the existence of the C_S sharing most of the compensated voltage, as shown in Figs. 6.23 and 6.24. It can be clearly seen U_S is in the same phase with I_S (Figs. 6.21 and 6.22).

Fig. 6.20 Simulation waveforms of proposed IUPQC in U_S, U_L

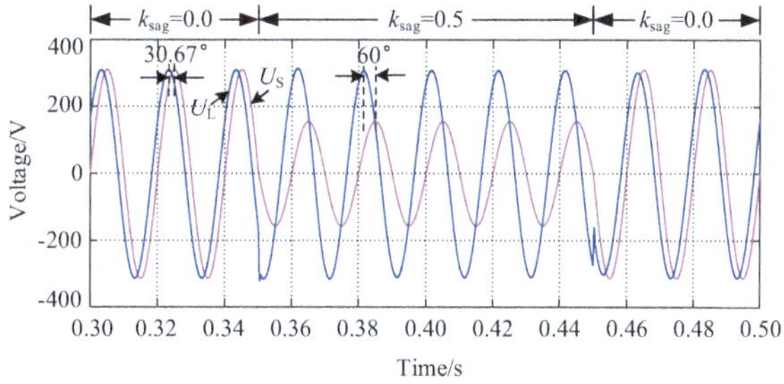

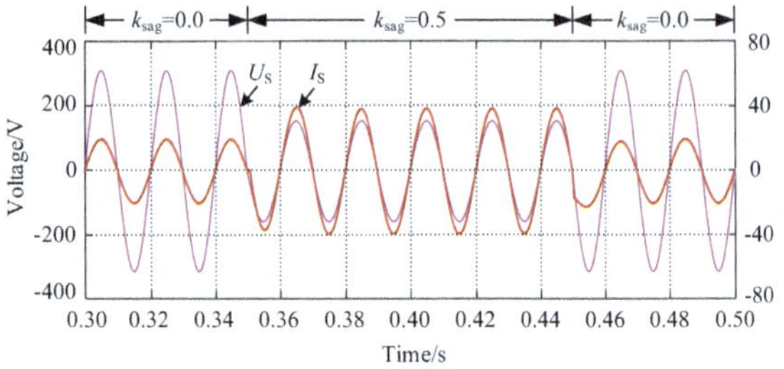

Fig. 6.21 Simulation waveforms of proposed IUPQC in U_S, I_S

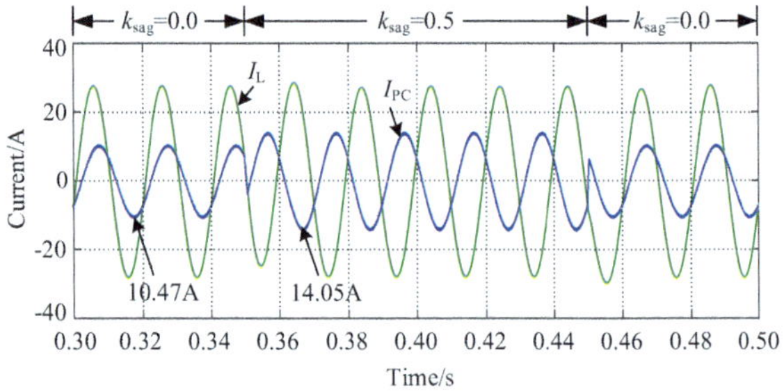

Fig. 6.22 Simulation waveforms of proposed IUPQC in I_L, I_{PC}

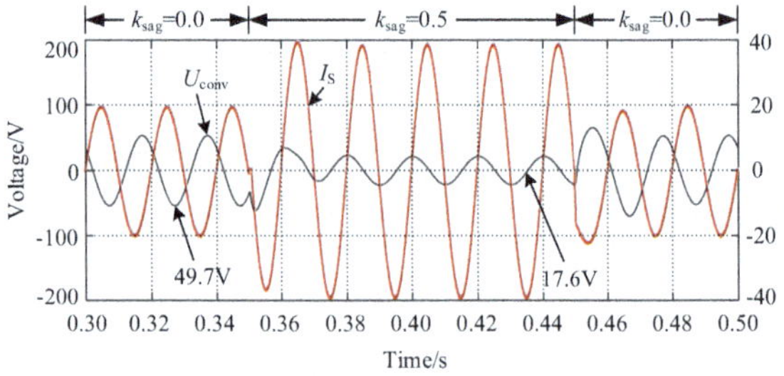

Fig. 6.23 Simulation waveforms of proposed IUPQC in U_{conv}, I_S

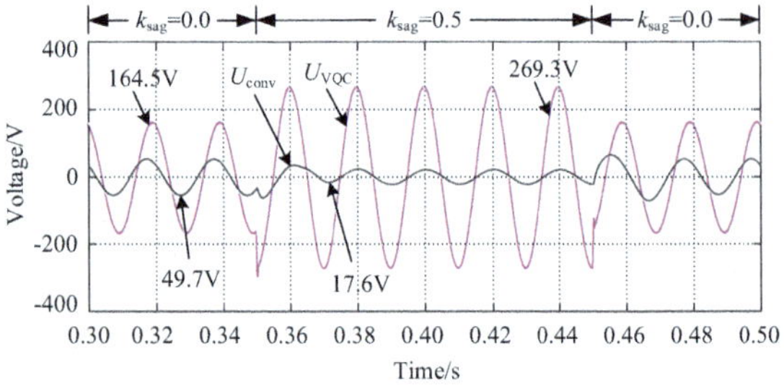

Fig. 6.24 Simulation waveforms of proposed IUPQC in U_{conv}, U_{VQC}

6.5.3 Performances of Proposed IUPQC Under 20% Voltage Swell Case

A voltage swell case with a depth of 20% is introduced during 0.35–0.45 s. As can be seen in Fig. 6.25, the amplitude of U_L is not affected. Although the U_{VQC} is very large (132.2 V, 0.425 p.u), the amplitude of U_{conv} still remains at a small value (80.4 V, 0.258p.u), as shown in Figs. 6.28 and 6.29. In addition, it can be clearly seen U_S is in the same phase with I_S (Figs. 6.26 and 6.27).

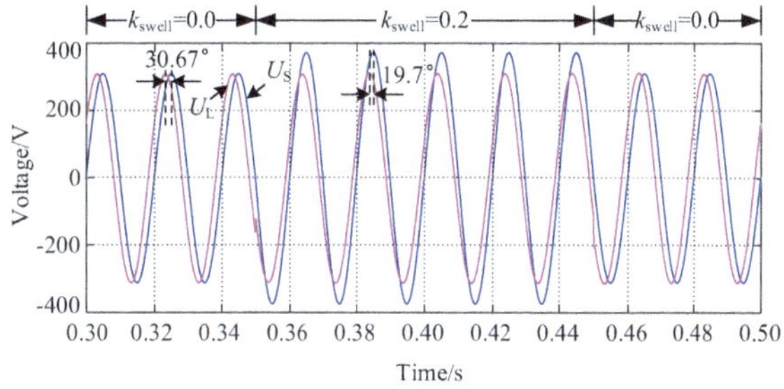

Fig. 6.25 Simulation waveforms of proposed IUPQC in U_S, U_L

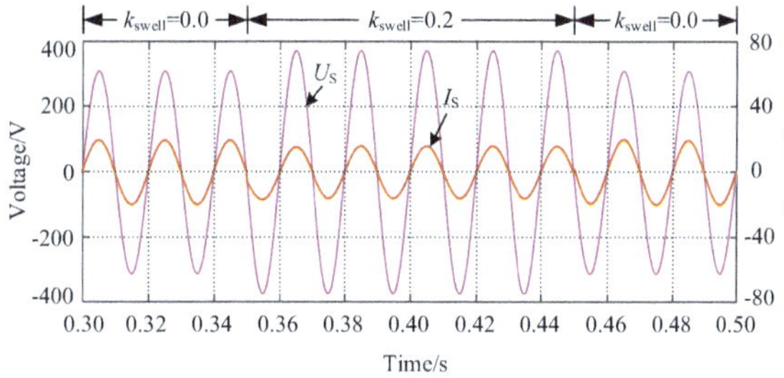

Fig. 6.26 Simulation waveforms of proposed IUPQC in U_S, I_S

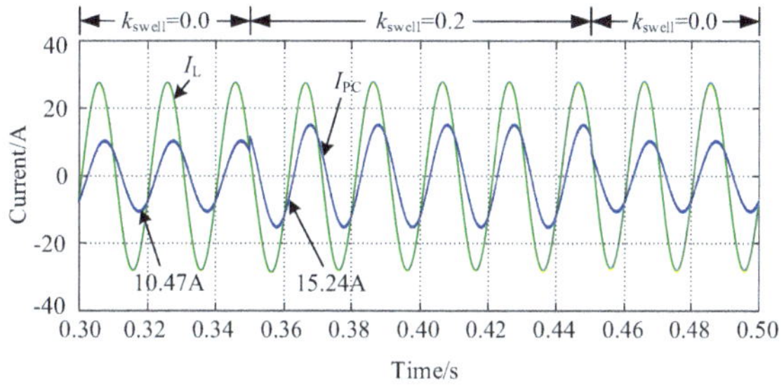

Fig. 6.27 Simulation waveforms of proposed IUPQC in I_L, I_{PC}

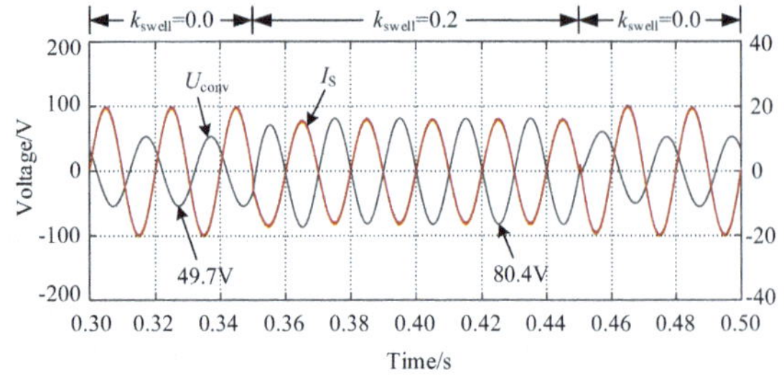

Fig. 6.28 Simulation waveforms of proposed IUPQC in U_{conv}, I_S

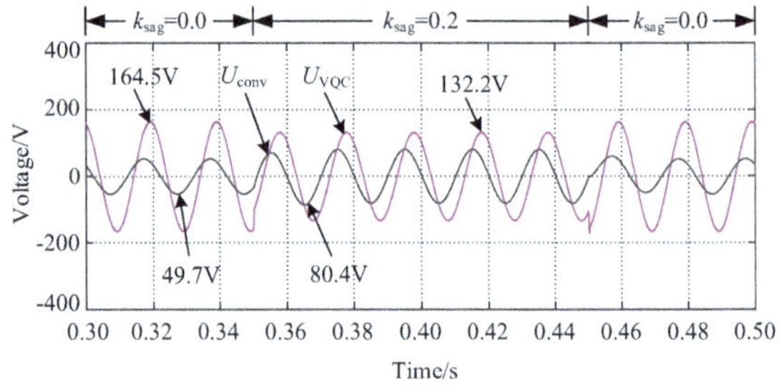

Fig. 6.29 Simulation waveforms of proposed IUPQC in U_{conv}, U_{VQC}

6.5.4 *Performances of Proposed IUPQC Under Voltage Distortion Case with Nonlinear Load*

In this scenario, the system voltage contains fifteen percent of the 5th and 7th harmonic components, respectively, as shown in Figs. 6.30 and 6.32. In addition, the harmonic distortion rate of load current is 20.46%. As shown in Fig. 6.31, the system current I_S basically contains only the power frequency component. In addition, as shown in Fig. 6.31, the load voltage U_L only contains the power frequency component (Figs. 6.33 and 6.34).

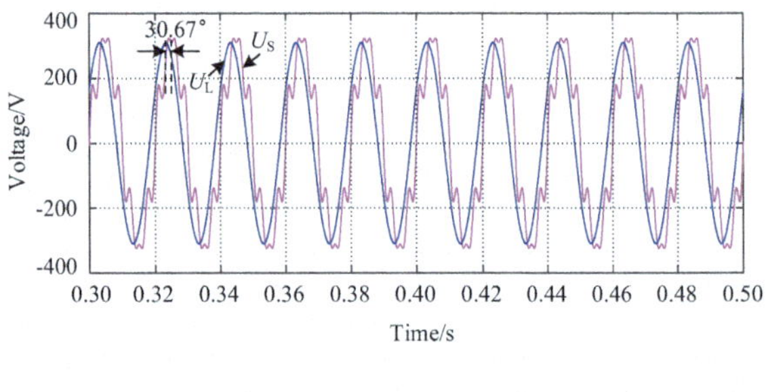

Fig. 6.30 Simulation waveforms of proposed IUPQC in U_S, U_L

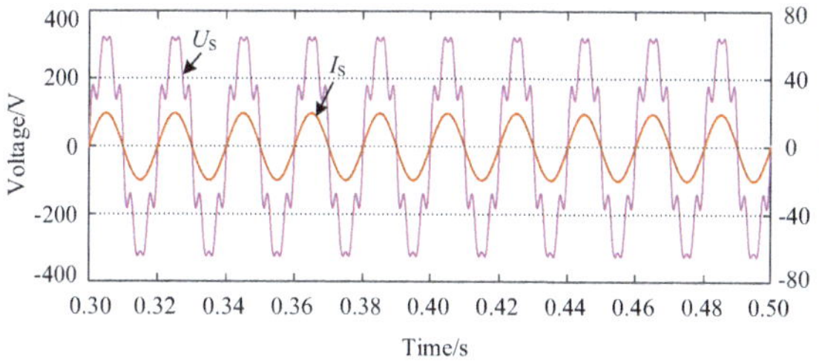

Fig. 6.31 Simulation waveforms of proposed IUPQC in U_S, I_S

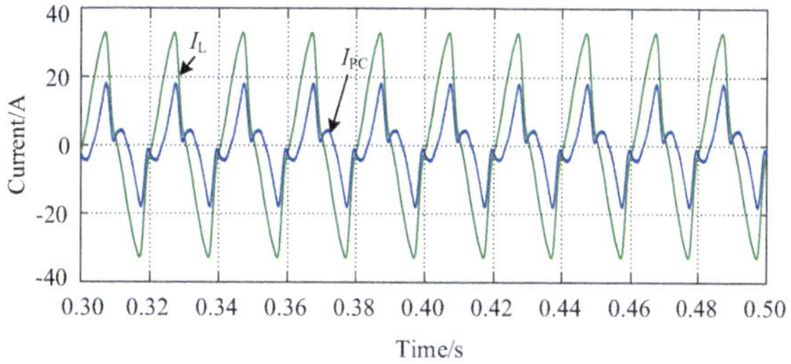

Fig. 6.32 Simulation waveforms of proposed IUPQC in I_L, I_{PC}

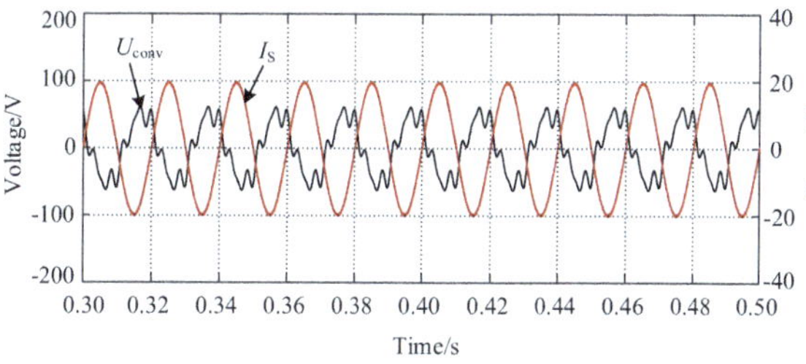

Fig. 6.33 Simulation waveforms of proposed IUPQC in U_{conv}, I_S

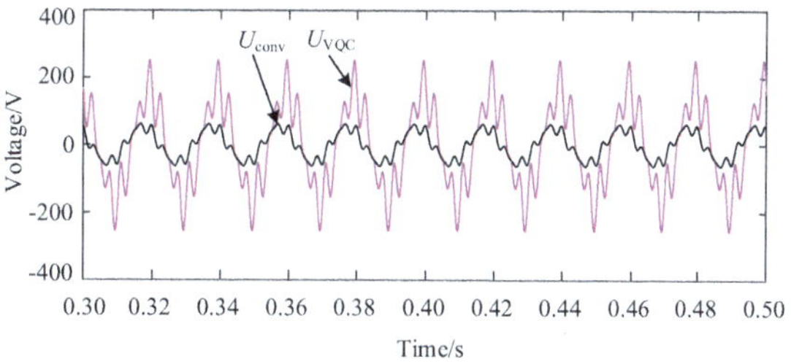

Fig. 6.34 Simulation waveforms of proposed IUPQC in U_{conv}, U_{VQC}

6.5.5 Performance Comparison Between Proposed IUPQC and Other UPQC

A comparative analysis of the injected voltage magnitude, the series transformer rating, and the converter rating requirements between the established hybrid UPQC [19, 20] and the proposed IUPQC is undertaken as follows. Drawing upon the data in Tables 6.2 and 6.3, when system voltage fluctuation incidents arise within a system voltage range of 0.5–1.2 per unit (p.u.), the peak combined converter rating for the proposed IUPQC and the conventional hybrid UPQC stands at 0.899 p.u. and 1.413 p.u., respectively. Consequently, the converter power rating in the IUPQC is reduced by 0.514 p.u. compared to the existing hybrid UPQC, underscoring the superiority of the IUPQC in terms of converter rating diminution, particularly in scenarios involving pronounced voltage sag events. However, when system voltage fluctuation issues manifest within a wider system voltage range of 0.5–1.5 p.u., the maximum converter ratings for the proposed IUPQC and the hybrid UPQC are 1.398 p.u. and 1.413 p.u., respectively. Under such circumstances, the benefits of the proposed strategy become less conspicuous, particularly in dealing with severe voltage swell conditions. This suggests that while the IUPQC offers notable advantages in handling voltage sags, its relative efficacy may not be as pronounced in managing voltage swells of greater magnitude.

In a broader perspective, given that professionals and academics tend to focus more on voltage sag instances ranging from 10 to 50% due to their prevalence in distribution networks [1], the methodology introduced herein holds a significant practical utility. Moreover, to further mitigate the power rating of the IUPQC specifically during voltage swell scenarios, there are two potential enhancements. One option involves augmenting the AC port of the IUPQC's parallel converter with an additional coupling capacitor; details regarding the integration of such a capacitor can be found in the literature [19, 20]. Alternatively, the constant series impedance (C_S) component on the series side of the IUPQC could be substituted with a tunable coupled LC

Table 6.2 Comparison of injected voltage and the series transformer rating between the proposed IUPQC and the existing hybrid UPQC

System voltage (p.u)	Injected voltage magnitudes (p.u)				Series transformer rating (p.u)	
	Existing hybrid UPQC		Proposed strategy		Existing hybrid UPQC	Proposed strategy
	U_{conv}	U_{VQC}	U_{conv}	U_{VQC}		
1.5 p.u	0.5	0.5	0.565	0.667	0.233	0.384
1.2 p.u	0.2	0.2	0.321	0.576	0.117	0.422
1.0 p.u	0	0.0	0.105	0.529	0.0	0.371
0.8 p.u	0.2	0.2	0.094	0.600	0.175	0.467
0.5 p.u	0.5	0.5	0.077	0.866	0.700	1.182

Table 6.3 Comparison of converter power rating between proposed IUPQC and the existing hybrid UPQC

System voltage (p.u)	Power rating of shunt converter (p.u)		Power rating of series converter (p.u)		Power rating of whole converter part (p.u)	
	Other hybrid UPQC	Proposed strategy	Other hybrid UPQC	Proposed strategy	Other hybrid UPQC	Proposed strategy
0.5–1.2	0.707	0.539	0.7	0.36	1.413	0.899
0.5–1.5	0.707	0.639	0.7	0.759	1.413	1.398

(TCLC) circuit featuring adaptable impedance characteristics. This dual approach not only caters to the reduction of power rating demands but also provides flexibility in addressing both voltage sags and swells, thereby enhancing the overall performance and efficiency of the IUPQC in dynamic grid environments. By tailoring the system to manage these critical fluctuations more effectively, the proposed solution contributes to the robustness and reliability of power distribution systems.

6.6 Experimental Verification

The laboratory-scale prototype of the IUPQC has been constructed to validate the effectiveness of its topology. The parameters of this prototype are detailed in Table 6.4. Figures 6.35, 6.36 and 6.37 demonstrate the response of the IUPQC to a 20% voltage sag event that occurs at time t_1.

Table 6.4 Parameters used in the experiment

Parameters	Values
System voltage amplitude and frequency	25 V, 50 Hz
Load parameters R_L and L_L	5.5 Ω, 17.5 mH
DC link capacitor C_{dc}	1640 μF
DC link voltage of IUPQC U_{dc}	20 V
LC filter inductor in VQC L_1	3 mH
LC filter capacitor in VQC C_1	60 μF
LC filter inductor in PC L_2	3 mH
Series coupled capacitor C_S	0.1mF
Switching frequency	10 kHz
k_{p1}, k_{i1}, k_{p2}, k_{i2}, k_{p3}, k_{i3}	0.25;0.1;0.12;0.1;2.2;0.1
Series transformer turn ratio	1:01
Parallel transformer turn ratio	2:01

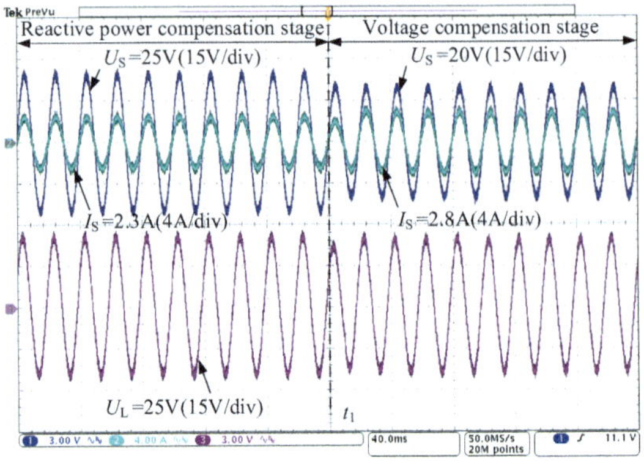

Fig. 6.35 Experimental performance of the IUPQC in U_S, U_L, and I_S

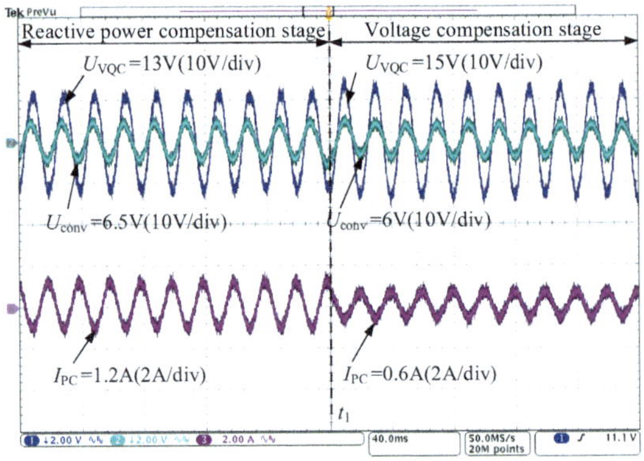

Fig. 6.36 Experimental performance of the IUPQC in U_{VQC}, U_{conv}, and I_{PC}

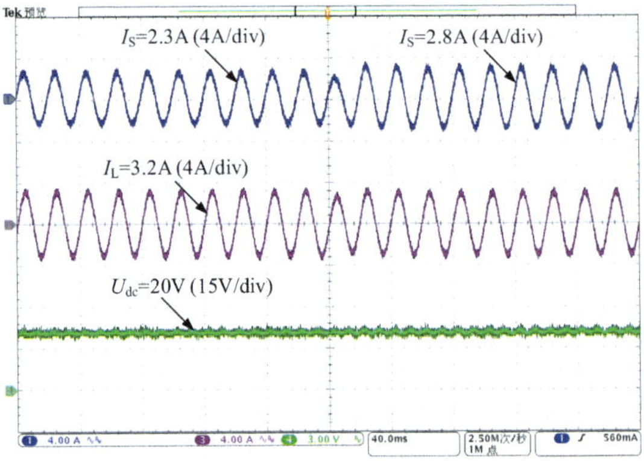

Fig. 6.37 Experimental performance of the IUPQC in U_{dc}, I_L, and I_S

Prior to t_1, the IUPQC operates in reactive power compensation mode, maintaining the grid voltage U_S at 1 p.u. (25 V). As evident from Fig. 6.35, U_S and I_S are in phase, indicating that the system is operating with unity power factor. U_L and U_S have identical amplitudes but differ in phase angles. Furthermore, Fig. 6.36 shows that while U_{VQC} reaches 13 V (0.52 p.u.), U_{conv} remains relatively low at 6.5 V (0.26 p.u.), demonstrating the effectiveness of the coupling capacitor in absorbing a significant portion of the voltage compensation. IPC is observed to be 1.2A (0.37 p.u.), which is less than the reactive component of I_L (0.707 p.u.), indicating that the parallel converter effectively handles the reactive power compensation.

Upon the occurrence of a 20% voltage sag at t_1, the IUPQC seamlessly transitions from reactive power compensation mode to voltage compensation mode. As illustrated in Fig. 6.35, the system continues to operate with unity power factor, ensuring stable operation. The compensated load voltage U_L remains at 25 V, effectively mitigating the voltage sag. From Fig. 6.36, it is evident that U_{conv} remains significantly lower than U_{VQC}, with values of 6 V (0.24 p.u.) and 15 V (0.6 p.u.), respectively, demonstrating the efficient voltage compensation capability of the IUPQC. Furthermore, I_{PC} is observed to be 0.6A (0.19 p.u.), which is less than the reactive component of the load current (0.707 p.u.), indicating that the parallel converter is adequately handling the compensation demands. Additionally, Fig. 6.37 shows that I_{PC} and U_{dc} remain stable, underscoring the robustness and stability of the IUPQC system during dynamic voltage disturbances.

References

1. R.A. Walling, R. Saint, R.C. Dugan, J. Burke, L.A. Kojovic, Summary of distributed resources impact on power delivery systems. IEEE Trans. Power Del. **23**(3), 1636–1644 (2008)
2. L. Campanhol, S. Silva, A. Oliveira, V. Bacon, Power flow and stability analyses of a multifunctional distributed generation system integrating a photovoltaic system with unified power quality conditioner. IEEE Trans. Power Electron. **34**(7), 6241–6256 (2019)
3. S. Dash, P. Ray, A new PV-open-UPQC configuration for voltage sensitive loads utilizing novel adaptive controllers. IEEE Trans. Ind. Inform. **17**(1), 421–429 (2021)
4. H. Hafezi, G. D'Antona, A. Dedè, D. Giustina, R. Faranda, G. Massa, Power quality conditioning in LV distribution networks: results by field demonstration. IEEE Trans. Smart Grid. **8**(1), 418–427 (2017)
5. B. França, L. Silva, M. Aredes, M. Aredes, An improved iUPQC controller to provide additional grid-voltage regulation as a STATCOM. IEEE Trans. Ind. Electron. **62**(3), 1345–1352 (2015)
6. V. Khadkikar, Enhancing electric power quality using UPQC: a comprehensive overview. IEEE Trans. Power Electron. **27**(5), 2284–2297 (2012)
7. B. Han, B. Bae, S. Baek, G. Jang, New Configuration of UPQC for Medium-voltage Application. IEEE Trans. Power Del. **21**(3), 1438–1444 (2006)
8. V. Khadkikar, A. Chandra, A.O. Barry, T.D. Nguyen, Power quality enhancement utilising single-phase unified power quality conditioner: digital signal processor-based experimental validation. IET Power Electron. **4**(3), 323–331 (2011)
9. W.C. Lee, D.M. Lee, T.K. Lee, New control scheme for a unified power-quality compensator-Q with minimum active power injection. IEEE Trans. Power Del. **25**(2), 1068–1076 (2010)

10. S. Devassy, B. Singh, Performance analysis of solar pv array and battery integrated unified power quality conditioner for microgrid systems. IEEE Trans. Ind. Electron. **68**(5), 4027–4035 (2021)
11. S. Khadem, M. Basu, M. Conlon, Intelligent islanding and seamless reconnection technique for microgrid with UPQC. IEEE J. Emerging Select. Topics Power Electron **3**(2), 483–492 (2015)
12. B.B. Ambati, V. Khadkikar, Optimal sizing of UPQC considering VA loading and maximum utilization of power-electronic converters. IEEE Trans. Power Del. **29**(3), 1490–1498 (2014)
13. A. Patel, H.D. Mathur, S. Bhanot, Enhancing VA sharing between the shunt and series APFs of UPQC with A modified SRF-PAC method. IET Power Electron. **13**(2), 275–285 (2020)
14. Y.Y. Kolhatkar, S.P. Das, Experimental investigation of a single-phase UPQC with minimum VA loading. IEEE Trans. Power Del. **22**(1), 373–380 (2007)
15. V. Khadkikar, A. Chandra, A new control philosophy for a unified power quality conditioner (UPQC) to coordinate load-reactive power demand between shunt and series inverters. IEEE Trans. Power Del. **23**(4), 2522–2534 (2008)
16. V. Khadkikar, A. Chandra, UPQC-S: a novel concept of simultaneous voltage sag/swell and load reactive power compensations utilizing series inverter of UPQC. IEEE Trans. Power Electron. **26**(9), 2414–2425 (2011)
17. S. Silva, L. Campanhol, G. Pelz, V. Souza, Comparative performance analysis involving a three-phase UPQC operating with conventional and dual/inverted power-line conditioning strategies. IEEE Trans. Power Electron. **35**(11), 11652–11665 (2020)
18. J. Ye, H. Gooi, F. Wu, Optimization of the size of UPQC system based on data-driven control design. IEEE Trans. Smart Grid **9**(9), 2999–3008 (2018)
19. S.B. Karanki, N. Geddada, M.K. Mishra, B.K. Kumar, A modified three-phase four-wire UPQC topology with reduced DC-link voltage rating. IEEE Trans. Ind. Electron. **60**(9), 3555–3566 (2013)
20. T. Koroglu, A. Tan, M. Savrun, M. Cuma, K. Bayindir, M. Tumay, Implementation of a novel hybrid UPQC topology endowed with an isolated bi-directional DC-DC converter at DC link. IEEE J. Emerging Select. Topics Power Electron. **8**(3), 2733–2746 (2019)
21. J. Wang, Kai Sun, H. Wu, J. Zhu, Y. Xing, Y. Li, Hybrid connected unified power quality conditioner integrating distributed generation with reduced power capacity and enhanced conversion efficiency. IEEE Trans. Indus. Electron. (2020). https://doi.org/10.1109/TIE.2020.3040687
22. Q. Guo, C. Tu, F. Jiang, R. Zhu and J. Gao, Improved dual-functional DVR with integrated auxiliary capacitor for capacity optimization. IEEE Trans. Ind. Electron. **68**(10), 3967–3975, 9755–9767 (2021)
23. B. Lu, Q. Guo, C. Tu, J. Gao, F. Jiang, F. Xiao, Capacity optimization scheme of UPQC system based on an improved topology. In: The 46th Annual Conference of the IEEE Industrial Electronics Society, Oct 2020, pp. 2457–2462
24. M. Kesler, E. Ozdemir, Synchronous-reference-frame-based control method for UPQC under unbalanced and distorted load conditions. IEEE Trans. Ind. Electron. **58**(9), 3967–3975 (2011)
25. Y. Lu, G. Xiao, X. Wang, F. Blaabjerg, D. Lu, Control strategy for single-phase transformerless three-leg unified power quality conditioner based on space vector modulation. IEEE Trans. Power Electron. **31**(4), 2840–2850 (2016)
26. V. Cheung, R. Yeung, H. Chung, A. Lo, W. Wu, A transformer-less unified power quality conditioner with fast dynamic control. IEEE Trans. Power Electron. **33**(5), 3926–3937 (2018)
27. R. Santos, J. Cunha, M. Mezaroba, A simplified control technique for a dual unified power quality conditioner. IEEE Trans. Ind. Electron. **6**(11), 5851–5860 (2014)
28. B. Bae, J. Jeong, J. Lee, B. Han, Novel sag detection method for lineinteractive dynamic voltage restorer. IEEE Trans. Power Del. **25**(2), 1210–1211 (2010)
29. C. Meyer, R.W. Doncker, Y.W. Li, F. Blaabjerg, Optimized control strategy for a medium-voltage DVR—theoretical investigations and experimental results. IEEE Trans. Power Electron. **23**(6), 2746–2754 (2008)
30. F.B. Ajaei, S. Afsharnia, A. Kahrobaeian, S. Farhangi, A fast and effective control scheme for the dynamic voltage restorer. IEEE Trans. Power Del. **26**(4), 2398–2406 (2010)

31. Q. Ye, R. Mo, H. Li, Impedance modeling and DC bus voltage stability assessment of a solid-state-transformer-enabled hybrid AC–DC grid considering bidirectional power flow. IEEE Trans. Ind. Electron. **67**(8), 6531–6541 (2020)
32. S. Biricik, H. Komurcugil, N.D. Tuyen, M. Basu, Protection of sensitive loads using sliding mode controlled three-phase DVR with adaptive notch filter. IEEE Trans. Ind. Electron. **66**(7), 5465–5475 (2019)

Chapter 7
Adaptive Soft Starter for Three-Phase Induction Motor Driving Device Using Dual-Functional Voltage Quality Controller

This section proposes an adaptive soft-starter control for induction motors using a DF-VQC. The proposed DF-VQC can operate in two modes: (i) the compensation mode for voltage sags and (ii) the virtual impedance mode for soft starting.

7.1 Introduction

Induction motors, as highlighted in [1], are pivotal components powering a myriad of operations across industries, manufacturing facilities, and agricultural settings. Ensuring the reliability and efficiency of these motors is paramount for maintaining product quality and the integrity of load-sensitive applications. Nonetheless, as noted in [2], the direct online starting of induction motors can pose challenges when they are powered from the point of common coupling (PCC). This is due to the voltage dips and current spikes that typically occur during the motor startup phase. Unregulated startup procedures can lead to the inadvertent tripping of overcurrent or undercurrent protective relays, potentially causing the motor to fail during the initiation sequence. Hence, as emphasized in [3], when conceptualizing a startup strategy for induction motors, engineers and researchers must meticulously address these issues to prevent such disruptions.

Various methodologies have been devised to tackle the challenges associated with the startup of induction motors, as documented in [4–8]. Among the commonly utilized apparatus for mitigating these issues are auto-transformers, line resistors, line reactors, wye-delta starters, and variable frequency drives (VFDs). More recently, soft starters have emerged as a viable solution, offering the capability to adjust the voltage supplied to motors through either mechanical means or power electronics. Thyristor-based soft starters, depicted in Fig. 7.1a, have gained widespread adoption in industrial settings due to their straightforward design and economic viability, as noted in [9]. They enable facile regulation of motor torque and acceleration during the

F. Jiang et al., *Advanced Voltage Quality Controller for New Power Distribution Systems*, https://doi.org/10.1007/978-981-97-9861-2_7

startup phase. Nevertheless, as pointed out in [10, 11], thyristor-based soft starters are not without their drawbacks, including suboptimal power factors and the introduction of harmonic distortions into the power system. Hence, there is a continuing demand for research and development aimed at refining soft-starting circuits to overcome these limitations and achieve superior performance. Innovations in this area seek to enhance the efficiency and reliability of motor startups while minimizing adverse effects on power quality. This pursuit is crucial for advancing the state of the art in motor control and ensuring the seamless operation of industrial equipment.

An additional impetus for investigating soft starters stems from their potential to augment the efficiency of induction motor drive systems, as highlighted in [12]. Within power systems, issues concerning power quality inevitably influence the

Fig. 7.1 **a** Industrial-type three-phase thyristorized soft starter and **b** a VQC for compensating voltage sags

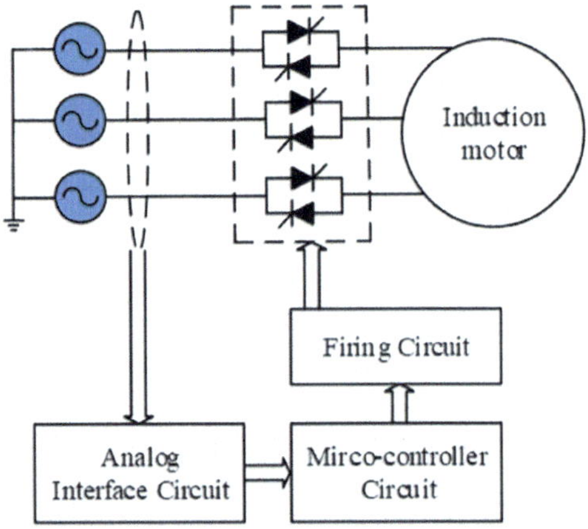

(a) Industrial-type three-phase thyristorised soft starter

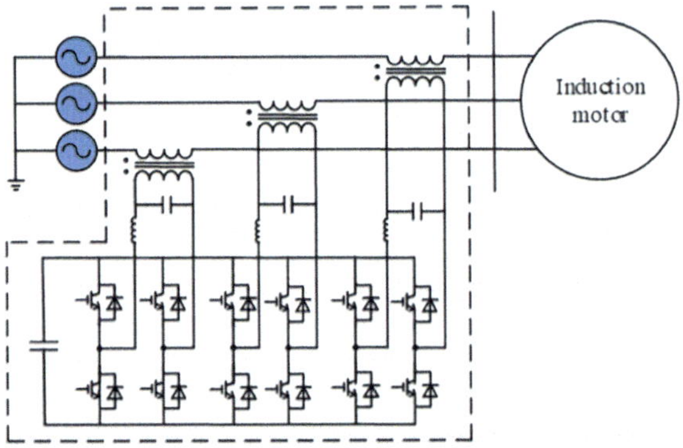

(b) a VQC for compensating voltage sags.

performance of induction motors, as elucidated in [13]. Despite ongoing efforts, these power quality challenges remain unresolved.

DVRs, depicted in Fig. 7.1b, are acknowledged as potent solutions for enhancing voltage quality. Academic research has explored the application of DVRs specifically for motor-driven scenarios. A dual feed-forward control scheme for DVRs was introduced in [14] to alleviate voltage dips triggered by the startup of high-power induction motors. Studies detailed in [15–17] corroborated the efficacy of DVRs in rectifying voltage sags resulting from short circuits and the startup of large induction motors within power systems through simulation outcomes.

However, these strategies predominantly concentrate on bolstering grid stability and overlook the specific requirements of induction motors, particularly those categorized as sensitive loads. Consequently, when deployed for motor loads, the DVR is primarily tasked with mitigating voltage sags on the supply side, rather than optimizing the steady-state operation of induction motors or diminishing the impact of startup currents. There exists a gap in addressing how to fine-tune the DVR's operation to concurrently ensure the robust performance of induction motors and minimize the disturbances caused by inrush currents during startup.

As illustrated in Fig. 7.1a, b, both thyristorized soft starters and DVRs, often referred to as VQCs, play critical roles in the operation of induction motors. The former is prevalently employed to facilitate the online startup of these motors, whereas the latter ensures a steady voltage at the motor terminals, thereby stabilizing the motor operation under varying grid conditions.

However, this dual-device configuration presents inefficiencies: during voltage compensation phases or motor startup, one device is actively engaged while the other remains dormant, leading to underutilization of resources. This redundancy not only escalates the overall system cost but also requires substantial physical space for installation. Additionally, the interplay between multiple devices can introduce complexities in motor performance, potentially impacting operational reliability and efficiency. Hence, the deployment of two distinct, single-purpose devices is suboptimal for induction motor setups. An integrated solution that consolidates the functionalities of both the thyristorized soft starter and the DVR into a unified system would address these issues by enhancing resource utilization, reducing costs, saving space, and simplifying the motor's operational characteristics. Such a multifunctional device would streamline the motor control process, making it a more desirable and practical choice for induction motor applications.

To tackle these challenges, integrating a single apparatus capable of performing multiple tasks is a viable approach. In light of this, this book introduces the DF-VQC, which represents an advancement over conventional thyristor-based soft starters and VQCs. The DF-VQC boasts not only high-efficiency soft-start capabilities but also the ability to counteract voltage fluctuations at the source. The subsequent sections of this book are structured as follows: Sect. 7.2 delves into the proposed topology and its dual functionality. Section 7.3 outlines an adaptive soft-starter control strategy tailored for the DF-VQC. Section 7.4 discusses a method for selecting optimal parameters for the DF-VQC. Section 7.5 presents both simulation and experimental findings, validating the performance of the proposed topologies.

7.2 DF-VQC Topology

Figure 7.2 illustrates that the DF-VQC is comprised of two principal elements: thyristor-controlled switches and series inverters. During routine operation, the bypass switch (designated as K) remains engaged, serving to minimize system losses. In scenarios where an induction motor initiates or when there is a dip in the source voltage, the bypass switch K is deactivated. This action facilitates both the soft starting of the motor and the compensation of voltage dips, ensuring a smooth startup process without undue stress on the electrical network.

The series inverter configuration includes a direct current (DC) capacitor (C_{dc}), a trio of three-phase H-bridge inverters (one dedicated to each phase) as detailed in Ref. [18], and a set of LC filters. The DC capacitor acts as an energy reservoir, supplying the necessary power to counteract voltage sags experienced on the utility grid. Each LC filter consists of inductors (represented as $L_f(a, b, c)$) and capacitors ($C_f(a, b, c)$), working together to smooth out the power delivery.

Notably, the DF-VQC design omits the use of transformers typically found in conventional VQC topologies, such as depicted in Fig. 7.1b. This omission, as discussed in Ref. [19], leads to reductions in cost, weight, and operational losses. Additionally, by assigning a distinct DC link to each single-phase inverter, as suggested in Ref. [20], the risk of short circuits is mitigated. The exclusion of the injection transformer further alleviates security concerns often associated with transformer-related issues, contributing to an overall safer and more efficient system design.

The equivalent circuit of an induction motor is shown in Fig. 7.3a and used to calculate the electrical parameters of the induction motors under different modes.

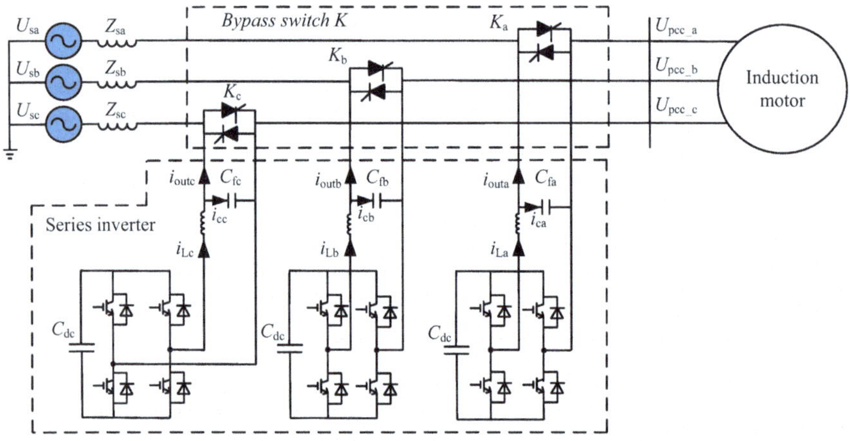

Fig. 7.2 Configuration of the proposed DF-VQC

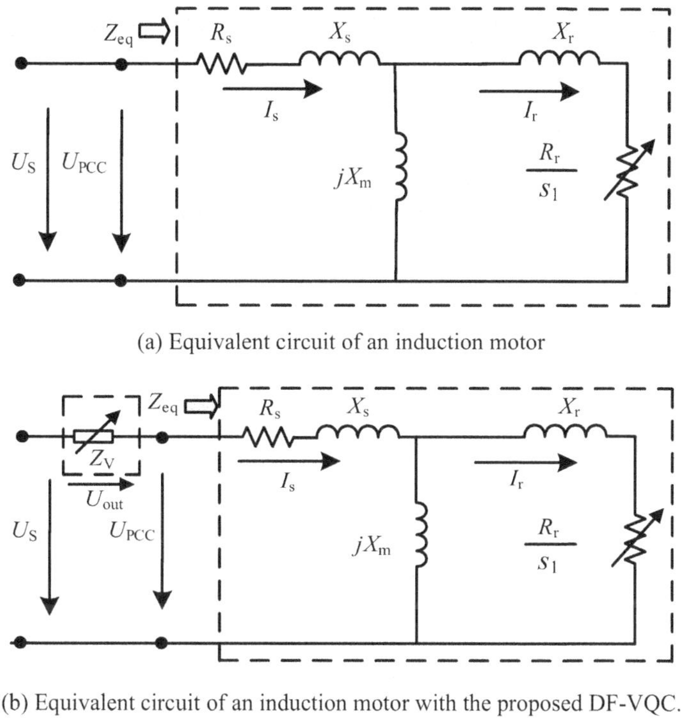

Fig. 7.3 **a** Equivalent circuit of an induction motor. **b** Equivalent circuit of an induction motor with the proposed DF-VQC

The equivalent impedance of the induction motor (Z_{eq}) is

$$Z_{eq} = (R_S + jX_S) + \frac{(jX_m)(R_r/s_1 + jX_r)}{(jX_m) + (R_r/s_1 + jX_r)}. \tag{7.1}$$

The stator side current (I_s) is

$$i_S = \frac{u_{pcc}}{Z_{eq}} = \frac{u_{pcc}}{(R_S + jX_S) + \frac{(jX_m)(R_r/s_1 + jX_r)}{(jX_m) + (R_r/s_1 + jX_r)}}. \tag{7.2}$$

The electromagnetic torque (T_{em}) is

$$T_{em} = \frac{P_{em}}{\Omega_1} = \frac{3I_r'^2 R_r}{\Omega_1 S_1} = \frac{3I_r'^2 R_r}{S_1} \cdot \frac{P}{4\pi f_1}, \tag{7.3}$$

where R_s and X_s are the resistance and leakage reactance of the stator, respectively; R_r and X_r are the resistance and leakage reactance of the rotor at the stator side, respectively; X_m is the magnetizing reactance; S_1 is the slip; I_s and I_r' are the stator and rotor currents at the stator side, respectively; U_{pcc} is the voltage per phase at the motor terminal; P_{em} is the electromagnetic power, and f_1 is the frequency. (7.1)–(7.3) indicate that the electromagnetic torque (T_{em}) can be expressed as a function of the motor terminal voltage per phase (U_{pcc}) as follows

$$T_{em} = \frac{3pR_f}{4\pi f_1 s_1} \cdot I_r'^2 = \frac{3pR_f}{4\pi f_1 s_1} \cdot U_{pcc}^2 \cdot |\frac{jXm}{\alpha + j\beta}|, \tag{7.4}$$

where p is the number of poles of the induction motor, and

$$\begin{cases} \alpha = R_S R_r / s_1 - X_s X_r - X_m (X_s + X_r) \\ \beta = R_S X_r + R_r X_s / s_1 + X_m (R_s + R_r / S_1) \end{cases}. \tag{7.5}$$

Another way to state T_{em} is

$$T_{em} = \frac{3pR_r}{4\pi f_1 s_1} \cdot |\frac{jX_m}{R_r/s_1 + j(X_r + X_m)}|^2 \cdot I_s^2. \tag{7.6}$$

Equation (7.6) shows that T_{em} can also be expressed as a function of stator current I_s. Therefore, the electromagnetic torque can be adjusted using (7.4) and (7.6) to control the speed of the induction motor by setting U_{pcc} and I_s. In other words, the induction motor can be soft started by controlling U_{pcc} and I_s. In this work, we can adopt the voltage-divider theorem to control U_{pcc} and I_s by inserting a virtual impedance into the line.

7.2.1 Voltage-Compensation Function of the DF-VQC

During regular operation, the DF-VQC proposed in this setup is effectively bypassed via switch K, allowing for direct power flow without interference. However, in scenarios where the source voltage exhibits imbalance or distortion, the DF-VQC springs into action, thus switching to voltage compensation operating mode.

In this mode, the DF-VQC meticulously introduces a corrective voltage, accurately calibrated in both amplitude and phase, into the circuit in series with the main line. This intervention is designed to counteract and mitigate any disturbances present in the utility grid voltage. Power for this corrective function is supplied by an integrated energy storage device, which ensures that adequate resources are available to compensate for voltage dips or sags experienced on the grid. This proactive approach to voltage regulation facilitates consistent power quality, even under adverse grid conditions. The energy storage device provides the required power to compensate for the voltage sags in the utility grid. The injected voltage is derived as follows:

$$u_{\mathrm{vqc}(a.b.c)} = u_{\mathrm{s}(a.b.c)} - u_{\mathrm{pcc}(a.b.c)} = \alpha u_{\mathrm{dc}} - u_{\mathrm{Lf}}, \tag{7.7}$$

where u_{dc} is the DC link voltage; α denotes the modulation index of the DF-VQC, and u_{Lf} is the voltage of the filter.

7.2.2 *Soft-Starting Function of the DF-VQC*

During induction motor startup, slip s_l is approximately equal to 1. (7.2) and (7.6) indicate that the starting current is derived by

$$i_{\mathrm{s\_t}} = \frac{u_{\mathrm{pcc}}}{(R_{\mathrm{S}} + jX_{\mathrm{S}}) + \frac{(jX_{\mathrm{m}})(R_{\mathrm{r}}+jX_{\mathrm{r}})}{(jX_{\mathrm{m}})+(R_{\mathrm{r}}+jX_{\mathrm{r}})}}. \tag{7.8}$$

Thus, the starting torque is given by

$$T_{\mathrm{em}} = \frac{3pR_{\mathrm{r}}}{4\pi f_1} \cdot \left|\frac{jX_{\mathrm{m}}}{R_{\mathrm{r}} + j(X_{\mathrm{r}} + X_{\mathrm{m}})}\right|^2 \cdot I_{\mathrm{s\_t}}^2, \tag{7.9}$$

Equations (7.8) and (7.9) show that $I_{\mathrm{s\_t}}$ is proportional to the PCC voltage U_{pcc} while T_{em} is proportional to the square of the starting current I_{st}. Hence, T_{em} is proportional to the square of U_{pcc}.

Virtual impedance is adopted to limit large starting currents and rotational speeds during starting, as shown in Fig. 7.3b. By adding a virtual impedance controller to the DF-VQC, the output impedance (Z_{V}) can be varied to adjust U_{pcc} at the motor terminal and $I_{\mathrm{s\_t}}$ during the starting transients. The starting current is given by

$$i'_{\mathrm{s\_t}} = \frac{u_{\mathrm{PCC}}}{Z_{\mathrm{V}} + Z_{\mathrm{eq}}} = \frac{u_{\mathrm{PCC}}}{Z_{\mathrm{V}} + (R_{\mathrm{S}} + jX_{\mathrm{S}}) + (jX_{\mathrm{m}})//(R_{\mathrm{r}} + jX_{\mathrm{r}})}, \tag{7.10}$$

Equations (7.8) and (7.10) show that the starting current is adjusted by controlling the virtual impedance.

7.3 Control Scheme and Parameter Selection

The DF-VQC may operate in different modes, i.e., the voltage compensation and virtual impedance modes, according to the state of the induction motor. In this section, the control scheme and the optimal parameter selection are explained in detail.

7.3.1 Control Scheme Design

(1) Fig. 7.4a depicts the control mechanism for adjusting the virtual impedance within the system. During standard operational scenarios, where the line current remains within acceptable limits, the virtual resistor remains dormant. In such cases, the DF-VQC functions primarily in the voltage compensation operating mode.

The virtual impedance (Z_V) is a crucial element in this system, and its adjustment is facilitated by the control scheme. The reference voltages for the voltage loop (u^*_{vqc}) and the power system (us_{ref}) are key inputs to the control process. Additionally, the currents flowing through the filter inductance (i_{Lf}), the distribution line (i_{vqc}), and the filter capacitor (i_c) are monitored and utilized in the control algorithm.

The output voltage of the series inverter, denoted as u_{vqc}, and the inverter's own output voltage, labeled as u_{inv}, are essential components of the control loop. These voltages, along with the current measurements and reference voltages, enable the system to dynamically adjust the virtual impedance as needed, maintaining optimal system performance under various operating conditions.

Given that the switching frequency of IGBT is high, we can use the instantaneous value instead of the mean value of the AC variable. From KCL and KVL,

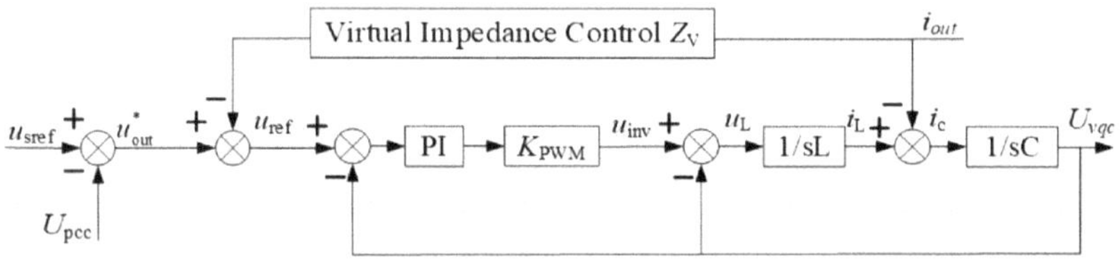

(a) Proposed DF-VQC control scheme in the stationary reference frame.

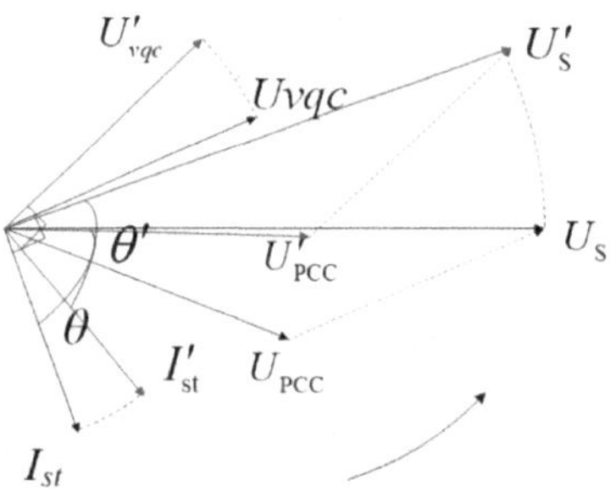

(b) Phase diagram of the voltage (U_{VQC}) across the virtual impedance.

Fig. 7.4 a Proposed DF-VQC control scheme in the stationary reference frame. **b** Phase diagram of the voltage (U_{VQC}) across the virtual impedance

$$L_{\mathrm{f}}\frac{di_{\mathrm{Lf}}}{dt} = u_{\mathrm{inv}} - u_{\mathrm{vqc}}. \tag{7.11}$$

$$C_{\mathrm{f}}\frac{du_{\mathrm{vqc}}}{dt} = i_{\mathrm{Cf}} = i_{\mathrm{Lf}} - i_{\mathrm{vqc}}, \tag{7.12}$$

$$u_{\mathrm{inv}} = L_f C_f \frac{d^2 u_{\mathrm{vqc}}}{dt} + L_f \frac{di_{\mathrm{vqc}}}{dt} + u_{\mathrm{vqc}}. \tag{7.13}$$

Figure 7.2 shows that we can track the output of the inverter arm. Thus,

$$u_{\mathrm{inv}} = u_{\mathrm{ref}} + k_{\mathrm{p}}(u_{\mathrm{ref}} - u_{\mathrm{vqc}}) + k_{\mathrm{I}} \int (u_{\mathrm{ref}} - u_{\mathrm{vqc}})dt. \tag{7.14}$$

The voltage across the filter capacitor is

$$\begin{aligned} u_{\mathrm{vqc}} &= \frac{(1+k_{\mathrm{p}})s + k_{\mathrm{I}}}{L_{\mathrm{f}}C_{\mathrm{f}}s^3 + (1+k_{\mathrm{p}})s + k_{\mathrm{I}}} u_{\mathrm{ref}} - \frac{L_{\mathrm{f}}s^3}{L_{\mathrm{f}}C_{\mathrm{f}}s^3 + (1+k_{\mathrm{p}})s + k_{\mathrm{I}}} \\ &= G_{\mathrm{V}}(s) \cdot u_{\mathrm{ref}} - Z_{\mathrm{inv}}(s) \cdot i_{\mathrm{vqc}} \\ &= G_{\mathrm{V}}(s) \cdot u_{\mathrm{ref}} - (R_{\mathrm{inv}} + jX_{\mathrm{inv}}) \cdot i_{\mathrm{vqc}}, \end{aligned} \tag{7.15}$$

where $G_{\mathrm{V}}(\mathrm{s})$ and $Z_{\mathrm{inv}}(\mathrm{s})$ are the voltage gain and output impedance of the inverter, respectively; k_{P} and k_{I} are the coefficients of the proportional and integral terms, respectively, and R_{inv} and X_{inv} are the equivalent output resistance and reactance.

When starting, the amplitude of the starting current is larger than I_{lim}, and the virtual impedance controller is activated. By incorporating an additional virtual impedance loop into the control system, the starting current can be limited to the desired value. From Fig. 7.4a, we know that

$$u_{\mathrm{ref}} = u_{\mathrm{vqc}}^{*} - Z_{\mathrm{V}}(s) i_{\mathrm{vqc}}. \tag{7.16}$$

The resistive elements of virtual impedance will cause the DF-VQC to absorb real power in Ref. [21]. Real power absorption is dangerous to the DF-VQC system and has a negative effect on motor starting. A better approach to this case would be to control the DF-VQC as a pure inductance (L_{V}) when starting $Z_{\mathrm{V}}(s) = j\omega L_{\mathrm{v}}$. By ignoring the influence of excitation impedance X_{m}, the system impedance angle is

$$\theta = \arccos \frac{(R_{\mathrm{s}} + R_{\mathrm{r}}/s) + R_{\mathrm{inv}}}{\sqrt{[(R_{\mathrm{s}} + R_{\mathrm{r}}/s) + R_{\mathrm{inv}}]^2 + (jX_{\mathrm{s}} + jX_{\mathrm{r}} + jX_{\mathrm{inv}} + jX_{\mathrm{V}})^2}}. \tag{7.17}$$

A phase diagram of the DF-VQC illustrating its ability to dynamically limit the starting current is shown in Fig. 7.4b.$u_{\mathrm{s}} = u_{\mathrm{Pcc}} + u_{\mathrm{out}}$ is the source voltage, and u_{out} is the compensation voltage injected orthogonally to the load current I_{st}. The system impedance angle (θ) follows (7.17) through adjustment of the virtual impedance Z_V. The injected voltage is

$$u_{vqc} = Z_v \cdot i_{s\_t}, \tag{7.18}$$

after starting the induction motors, the rotor speed increases gradually, and slip ratio s_1 decreases. Thus, $(1 - s_1)R_r/s_1$ will gradually increase. (7.10) indicates that the virtual impedance Z_V can be dynamically adjusted when starting to limit starting current $I'_{s,t}$. . We can even achieve $i_{s\_t} = I'_{s,t}$ when starting, as shown in Fig. 7.4b.

(2) Adaptive soft-starter control: If the virtual impedance is connected in series with the distribution line, the terminal voltages (u_{pcc}) can be expressed as

$$u_{pcc} = (1 - \frac{Z_V i_{vqc}}{u_s})u_s = a \cdot u_s, \tag{7.19}$$

where a is the ratio between u_{pcc} and u_s, the soft-starter current is a times less than the direct starting current, and the soft-starter torque is the square of a times less than the direct starting torque. Thus, $I'_{s,t}$ and T'_{em} are verified for different values of a, e.g., $a = 1/6$, and the influence of Z_{inv} is ignored

$$a\sqrt{(R_s + R_r/s_1)^2 + (jX_s + jX_r + jX_v)^2} = \sqrt{(R_s + R_r/s_1)^2 + (jX_s + jX_r)^2}, \tag{7.20}$$

and then

$$Z_{V0} = \sqrt{(1 - a^2)(R_s + R_r/s_1) + (jX_s + jX_r)^2} - (jX_s + jX_r), \tag{7.21}$$

where Z_{V0} is the initial value of the virtual impedance.

The dynamic adjustment process of Z_V is shown in Fig. 7.5a. The dynamic processes can be derived as

$$Z_V = Z_{V0} - \sum_{i=1}^{T} k \cdot J_{V0\_i}, \tag{7.22}$$

where T is the operating period of the virtual impedance, and J_{V0} is the correction factor of the virtual impedance. k should be

$$k = \begin{cases} 1, & \Delta I > 0 \\ 0, & \Delta I < 0 \end{cases}, \tag{7.23}$$

where $\Delta I = (I_{n+1} - I_n)$. Hence, the starting current can be limited to a reasonable value.

The illustration in Fig. 7.5b depicts the adaptive virtual impedance control system in operation. In this setup, I_{rms} represents the root mean square value of the stator current, while I_{ref} denotes the predetermined reference limit for the current. The system computes the difference between I_{rms} and I_{ref}, denoted as ΔI, and forwards this value to a switching mechanism.

Fig. 7.5 **a** Dynamic adjustment of X_V and J_{V0} as the starting current changes. **b** DF-VQC control scheme

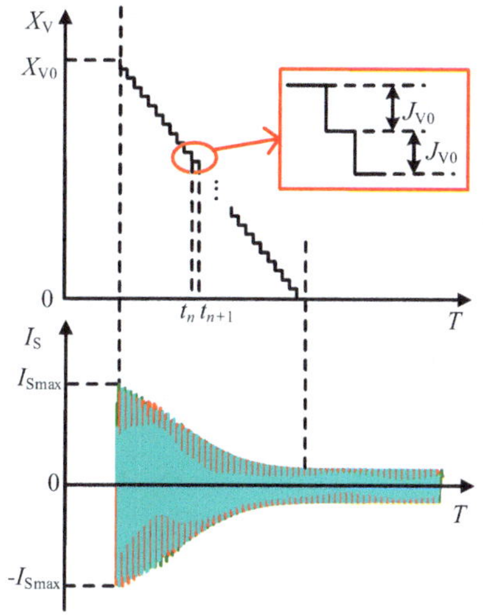

(a) Dynamic adjustment of X_V and J_{V0} as the starting current changes.

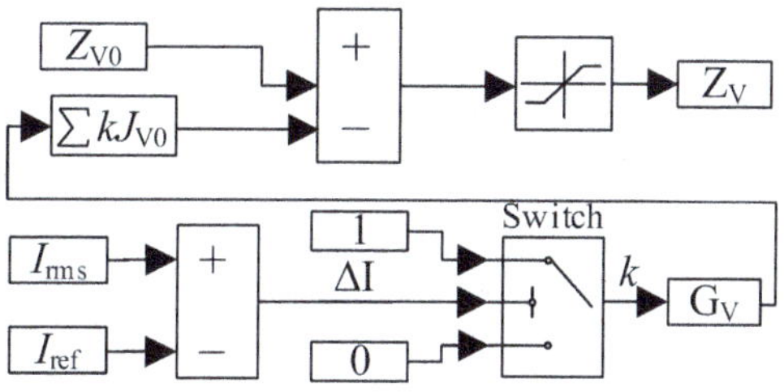

(b) DF-VQC control scheme.

If ΔI is less than 0, indicating that the stator current is below its reference limit, the switch outputs a signal of 0. Consequently, the virtual impedance retains its fixed value, ensuring that the stator terminal voltage (Upcc) also remains constant.

On the other hand, if ΔI is greater than 0, suggesting that the stator current has exceeded its reference limit, the switch outputs a signal of 1. This triggers an adjustment to JV_0, resulting in a decrease in the virtual impedance. Consequently, the stator terminal voltage (Upcc) increases, allowing for more effective current regulation and ensuring that the stator current does not exceed its safe operating limits.

7.3.2 Optimal Parameter Selection

(1) Virtual impedance value design: Using (7.15) and (7.16), the output impedance of the inverter (Z_{out}) after incorporating an additional virtual impedance is derived as

$$Z_{out}(s) = G_V(s) \cdot Z_V(s) + Z_{inv}(s). \tag{7.24}$$

Then, the stator current is expressed as

$$I_S(s) = \frac{U_S(s)}{G_V(s)Z_V(s) + Z_{inv}(s) + Z_{eq}(s)}. \tag{7.25}$$

To limit the first peak of the starting current, the initial output impedance L_{v0} should be set such that $I_{st} < I_{ref}$. Thus,

$$X_{V0} \geq \frac{a_1 s^4 + a_2 s^3 + a_3 s^2 + a_4 s^1 + a_5}{b_1 s + b_2}$$
$$\begin{cases} a_1 = -LC(X_s + X_r) \\ a_2 = LC(U(s)/I_{ref} - R_s - R_f) \\ a_3 = -L - (1 + k_p)(X_s + X_r) \\ a_4 = (U(s)/I_{ref} - R_s - R_f)(1 + k_p) - (X_s + X_r)k_I \\ a_5 = (U(s)/I_{ref} - R_s - R_f)k_I \\ b_1 = 1 + k_p \\ b_2 = k_I \end{cases} \tag{7.26}$$

(2) Switch design: The DF-VQC can bear a maximum current i_{max} because the starting current will not increase until the virtual impedance is in series with the distribution line. Thus, we have

$$i_{max} \leq \min(i_L, i_{IGBT}, i_D), \tag{7.27}$$

where i_L, i_{IGBT}, and i_D denote the maximum permissible currents of the filter inductor, IGBTs, and diodes, respectively. In extreme circumstances (e.g., short-circuit faults), the DF-VQC should withstand the supply voltage u_S, i.e.,

$$u_s \leq \min(u_L, u_{IGBT}, u_D), \tag{7.28}$$

where u_{IGBT}, u_D, and u_{DC} represent the rated blocking voltages of the IGBTs, diodes, and DC-side capacitor, respectively.

(3) LC design: when the DF-VQC operates in voltage compensation operating mode, the *LC* is designed to filter switching harmonics due to the IGBTs. L_f and

C_f are calculated through the method presented in reference [22] in the compensation voltage mode. The thermal stability of L_f and C_f should be considered during the fault and can be expressed as

$$Q_{Lf} \geq \int_{t_{starting}}^{t_{finish}} (ki_L)^2 dt, \tag{7.29}$$

where Q_{Lf} indicates the thermal stability of L_f during the fault, and $t_{starting}$ and t_{finish} denote the times at which the motor starts and the starting transient finishes, respectively. Considering that C_f can be easily damaged by overvoltage, u_{cf} may be expressed as

$$u_{Cf} \geq k_1 u_s, \tag{7.30}$$

where k_1 is the safety coefficient of C_f, $k_1 = 1.2$, and u_{cf} is the maximum withstanding voltage of the filter capacitor. Moreover, in industrial applications, a snubber circuit and varistors are designed to protect the filter capacitor in Refs. [8, 23].

(4) DC link design: The DF-VQC needs active power to restore the voltage waveform during the supply voltage sag, which can be supplied from an energy storage device, auxiliary supply or a line-connected shunt converter. Varistors are designed to protect the DC link capacitor during faults. The threshold value of the DC link protection voltage, U_{dc_th}, is set to 1.25 U_{dc}.

7.4 Simulation Results

In this section, simulation results are presented to show the feasibility of the proposed DF-VQC. Simulation of the circuit shown in Fig. 7.2 is performed using MATLAB/ Simulink software.

7.4.1 Soft-Start Function of the DF-VQC

Figure 7.6 illustrates the simulation outcomes pertaining to the motor current and rotor velocity across varying operational scenarios. As depicted in Fig. 7.6a, direct startup engenders a substantial inrush current peak, reaching approximately 1200 Amperes, which represents a staggering six-fold amplification over the rated current's apex. This significant surge poses a considerable risk to power electronics, potentially precipitating damage. The motor attains its operational status within a span of 0.9 s, as observed in the same figure.

In contrast, Fig. 7.6b shows the effects of using a thyristor soft starter. Here, the current is capably restrained to thrice the rated current magnitude, elongating the startup duration to 1.8 s. However, from the magnification part of Fig. 7.6b, it

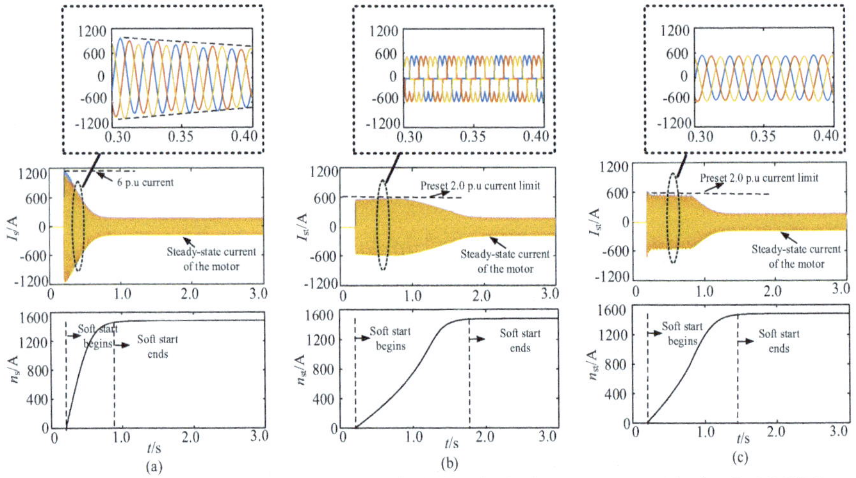

(a) Direct online starting mode (b) using the thyristorised soft starte (c) using the DF-VQC.

Fig. 7.6 Simulation results for soft-start comparison

is obvious that there is an obvious harmonic content in the starting current. This undesirable characteristic could jeopardize the integrity and stability of the electrical network, compromising its overall security.

Figure 7.6c showcases the distinct advantages of employing the proposed DF-VQC. Similar to thyristor soft start, the current is maintained at three times the rated current. Notably, the motor experiences a gentle acceleration phase prior to settling into a steady-state speed, achieving full startup in 1.5 s. The magnified section of Fig. 7.6c reveals that the startup current retains a sinusoidal profile, indicative of reduced harmonic distortion. Consequently, the DF-VQC showcases commendable operational attributes, enabling an efficient and smooth soft-start functionality.

7.4.2 Dynamic Switching Behavior Between the Two Functions

To validate the seamless transition between the two operational modes, switching simulations were conducted and their outcomes are presented in Fig. 7.7. As depicted in Fig. 7.7a, the soft start commences at t_1 and terminates at t_2. During this interval, the motor current is effectively regulated to the desired level by the thyristor-based soft starter. However, in the event of a 20% voltage sag occurring between t_3 and t_4, the thyristor-based starter is unable to compensate for the voltage drop, resulting in a decrease in the motor stator current. The magnified view in Fig. 7.7b reveals noticeable oscillations in the rotor speed.

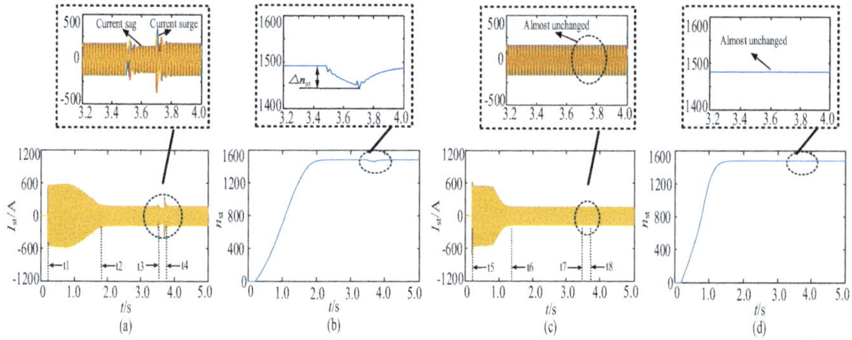

(a) Motor current with the thyristorised soft starter, (b) motor speed with the thyristorised soft starter, (c) motor current with the DF-VQC and (d) motor speed with the DF-VQC.

Fig. 7.7 Simulation results of switching between two function using different devices

In contrast, Fig. 7.7c, d demonstrates that the DF-VQC effectively compensates for the voltage sag from t_7 to t_8. During this period, the motor stator current remains relatively stable, and the rotor speed remains unaffected, showcasing the DF-VQC's superior dynamic response. Furthermore, the DF-VQC achieves better damping of the motor speed and a smoother power transition compared to the thyristor-based soft starter.

7.5 Experiment and Analysis

7.5.1 Soft-Start Function

Figure 7.8 illustrates the outcomes of practical tests conducted on the motor's current and rotor velocity under various control methodologies. In part (a) of Fig. 7.8, it is evident that direct starting results in a surge, causing the current to escalate to multiples far beyond the nominal current levels.

In contrast, Fig. 7.8b exhibits the effects of utilizing a thyristor-based soft starter, where the current is restrained to approximately three times the standard operating current. However, upon closer inspection within the magnified portion of Fig. 7.8b, one can observe the presence of significant harmonic content in the startup current.

Figure 7.8c portrays the operational scenario when employing the suggested soft starter, which also caps the startup current at around three times the rated current. Notably, the detailed view in Fig. 7.8c reveals that the startup current closely mirrors a sinusoidal waveform, indicating superior current quality.

These findings affirm that the proposed DF-VQC boasts commendable operational attributes and delivers a highly efficient soft-start capability. The consistency between the experimental data depicted in Fig. 7.8a–c and the corresponding simulation outcomes underscores the validity and reliability of the DF-VQC's performance.

Fig. 7.8 Experimental results of different controls

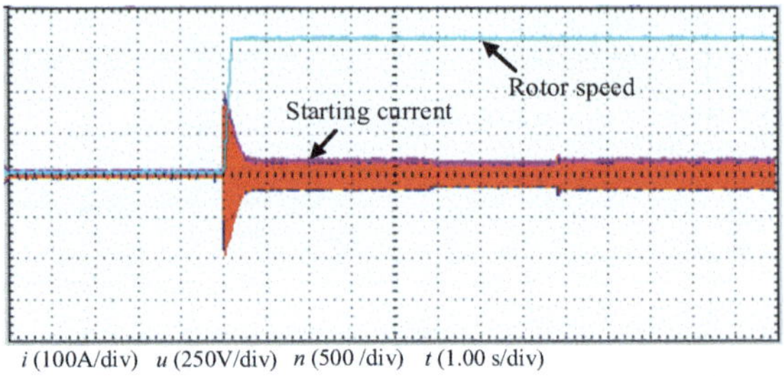

(a) Direct starting

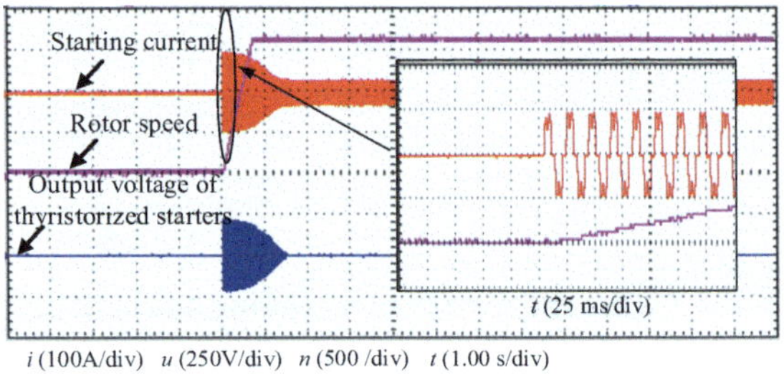

(b) with a thyristorised soft starter

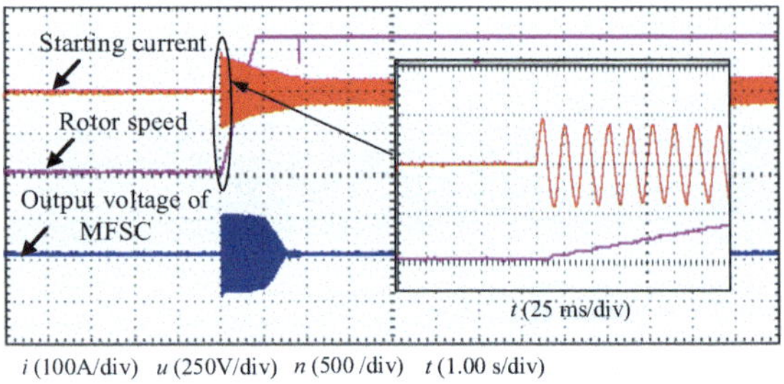

(c) with the proposed DF-VQC.

7.5.2 *Dynamic Switching Behavior Between the Two Functions*

The dynamic performance of the DF-VQC showed in Fig. 7.9 as it transitions between its two primary functions. The soft-start sequence initiates at t_1 and concludes at t_2, during which period the motor current, aided by the implemented soft starter, is effectively constrained to the predetermined current limit. Subsequently, when a voltage sag of 20% depth arises between t_3 and t_4, the DF-VQC swiftly injects a compensatory voltage, enabling the load current to remain stable. The detailed view of the figure underscores the DF-VQC's impressive dynamic responsiveness. Importantly, throughout this entire sequence, the rotor speed should remain unaffected, ensuring seamless operation.

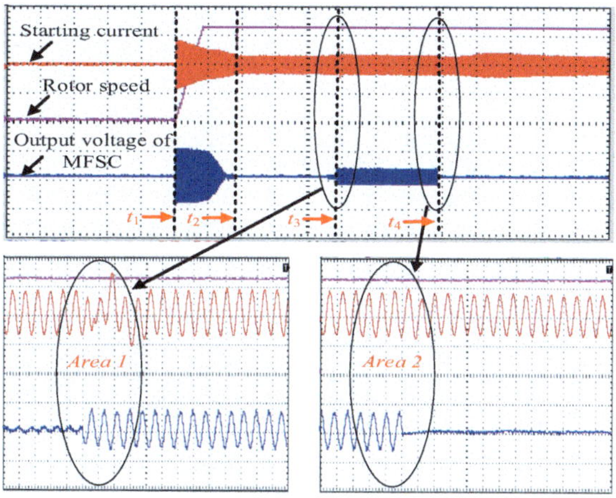

Fig. 7.9 Experimental results of switching between the two functions using the DF-VQC

References

1. C.C. Yeh, N.A.O. Demerdash, Fault-tolerant soft starter control of induction motors with reduced transient torque pulsations. IEEE Trans. Energy Convers. **24**(4), 848–859 (2009)
2. G. Zenginobuz, I. Cadirci, M. Ermis et al., Performance optimization of induction motors during voltage-controlled soft starting. IEEE Trans. Energy Convers. **19**(2), 278–288 (2004)
3. A. Anwar, Y. Zhang, C.W. Brice et al., Soft reclosing of an industrial power network using an inverter-controlled energy-storage system. IEEE Trans. Power Del. **29**(3), 1111–1119 (2014)
4. F. Endrejat, P. Pillay, The soft starters: adjustable speed systems for multiple MW rated motors. IEEE Trans. Appl. Mag. **14**(6), 27–37 (2008)
5. F.J.T.E. Ferreira, M.V. Cistelecan, A.T. De Almeida, Comparison of different tapped windings for flux adjustment in induction motors. IEEE Trans. Energy Convers. **29**(2), 375–391 (2014)
6. A. Gastli, M.M. Ahmed, ANN-based soft starting of voltage-controlled-fed IM drive system. IEEE Trans. Energy Convers. **20**(3), 497–503 (2005)
7. S. Mestry, P. Kushe, S. Kelkar, A. Patil, Smooth starting of induction motor during open circuit and short circuit condition, in *2018 Second International Conference on Inventive Communication and Computational Technologies (ICICCT)* (2018), pp. 1118–1122
8. J.A. Corral-Hernandez, J.A. Antonino-Daviu, Thorough validation of a rotor fault diagnosis methodology in laboratory and field soft-started induction motors. Chin. J. Electr. Eng. **4**(3), 66–72 (2018)
9. A R N M Reaz Ul Haque, Leng S, Smith I, et al.: "Performance analysis of grid connected induction motor using floating H-bridge converter," IEEE Applied Power Electronics Conference and Exposition., 2017, 3163–3170.
10. T.M. Rowan, T.A. Lipo, A quantitative analysis of induction motor performance improvement by SCR voltage control. IEEE Trans. Ind. Appl. **19**(4), 545–553 (1983)
11. M.G. Solveson, B. Mirafzal, N.A.O. Demerdash, Soft-started induction motor modeling and heating issues for different starting profiles using a flux linkage ABC frame of reference. IEEE Trans. Ind. Appl. **42**(4), 973–982 (2006)
12. K. Sundareswaran, B.M. Jos, Development and analysis of novel soft-starter energ-saver topology for delta-connected induction motors. IEE Proc.-Electr. Power Appl., 2005, 152, (4), pp. 1486–1497
13. J.C. Gomez, M.M. Morcos, C.A. Reineri et al., Behavior of induction motor due to voltage sags and short interruptions. IEEE Trans. Power. Del. **17**(2), 434–440 (2002)

14. J. Zhou, H. Zhou, Z. Qi, The study on a dual-feed-forward control of DVR to mitigate the impact of voltage sags caused by induction motor starting, in *Proceedings of International Conference on Electrical Machines and Systems*, Oct. 2008, pp. 1497–1500
15. F.M. Mahdianpoor, R.A. Hosshmand, M. Ataei, A new approach to multifunctional dynamic voltage restorer implementation for emergency control in distribution systems. IEEE Trans. Power. Del. **26**(2), 882–890 (2011)
16. R. Garcia-valle, L.C.P. da Silva, Z. Xu et al., Smart demand for improving short-term voltage control on distribution networks. IET Gener. Transm. Distrib. **3**(8), 724–732 (2009)
17. F. Jiang, C. Tu, Q. Guo, et al., Dual-functional dynamic voltage restorer to limit fault current. IEEE Trans. Ind. Electron. **6**(7), 5300–5309 (2019)
18. M. Farhadi-Kangarlu, E. Babaei, F. Blaabjerg, A comprehensive review of dynamic voltage restorers. Int. J. Electr. Power Energy Syst. **92**(11), 136–155 (2017)
19. W. Santos, E. Silva, C.B. Jacobina et al., The transformerless single-phase universal active power filter for harmonic and reactive power compensation. IEEE Trans. Power Electron. **29**(7), 3563–3572 (2014)
20. B.H. Li, S.S. Choi, D.M. Vilathgamuwa, Transformerless dynamic voltage restorer. IEE Proc.-Gener. Transm. Distrib. **149**(3), 263–273 (2002)
21. D.M. Vilathgamuwa, P.C. Loh, Y. Li, Protection of microgrids during utility voltage sags. IEEE Trans. Indus. Elect. **53**(5), 1427–1436 (2006)
22. F. Jiang, C. Tu, Z. Shuai et al., Multilevel cascaded-type dynamic voltage restorer with fault current limiting function. IEEE Trans. Power Del. **31**(3), 1261–1269 (2016)
23. L.A. Moran, I. Pastorini, J. Dixon et al., A fault protection scheme for series active power filters. IEEE Trans. Power Electron. **14**(5), 928–938 (1999)

Chapter 8
Conclusions

This book focuses on the urgent demand for high-quality and efficient development of active distribution networks, and effectively address the bottleneck issues such as single functionality, low equipment utilization, and poor reliability encountered in the practical process of the series-connected voltage quality controllers. It primarily conducts research on the expansion of functions and performance improvement of series-connected voltage quality controllers. Based on this, the book summarizes some achievements and experiences of the authors in the new topology, new technology, new methods, and engineering application cases of advanced series-connected voltage quality controllers, including:

(1) Elaboration on the working principle of typical series-connected voltage quality controllers, detailed analysis of the influencing factors on the output capability of series-connected voltage quality controllers, and formulation of smooth start-stop control strategies and high-performance output methods based on quantitative characterization of their output boundaries.
(2) Proposal of a method for constructing multifunctional series-connected voltage quality controllers with fault current-limiting capability. It constructs multifunctional series-connected voltage quality controllers based on filtering inductance reuse and DC-side leakage current branch reuse. The proposed optimized operation method ensures effective operation in different modes and enables flexible switching between different modes.
(3) Capacitively Coupled VQC and DF-VQC have been extended for capacity optimization in UPQC and motor starting process, respectively. Then, the power rating of IUPQC's power electronic module can be significantly reduced, the overcurrent during the motor starting process can also be effectively reduced as well.

The research of this book aims to arouse the attention of relevant researchers to the advanced voltage quality controller, and hope to provide useful reference for relevant scholars, engineers, and equipment manufacturers that are interested in

F. Jiang et al., *Advanced Voltage Quality Controller for New Power Distribution Systems*, https://doi.org/10.1007/978-981-97-9861-2_8

voltage quality controller and multifunctional composite system at home and abroad. This book can be used as a textbook for learning and studying smart grids, as well as a manual for solving practical engineering problems, and can enlighten the research on cutting-edge technologies.

The manufacturer's authorised representative in the EU is Springer Nature Customer Service Centre GmbH, Europaplatz 3, 69115 Heidelberg, Germany. If you have any concerns regarding our products, please contact ProductSafety@springernature.com

Printed and bound by CPI Group (UK) Ltd, Croydon, CR0 4YY
01/10/2026
02225980-0013